The 4D Spiral Spacetimes
Toryx & Helyx –
Prime Elements of the Multiverse

First Edition

by
Vladimir B. Ginzburg

Cover by Eugene B. Ginzburg

Helicola Press
Division of IRMC, Inc.
612 Driftwood Drive
Pittsburgh, Pennsylvania, 15238

The 4D Spiral Spacetimes
Toryx & Helyx –
Prime Elements of the Multiverse

First Edition

Published by
Helicola Press
Division of IRMC, Inc.
612 Driftwood Drive
Pittsburgh, Pennsylvania
USA

Printed in the United States of America

ISBN: 978-0-9671432-9-3 (Perfect bound)

Current printing (last digit)
10 9 8 7 6 5 4 3 2 1

Contents

My special thanks go to the members of my family: to my still young, gorgeous, friendly and energetic grandson Alex, to my beautiful and talented grandson Asher; to my daughter Ellen with whom I brainstormed some ideas related to my theory and who proofread Summary of this book; to my son Gene whose comments about a multi-level universe made in December of 1992 had triggered my interest in the development of the idea described in this book, and who also designed covers of all my books and helped me to produce them; to my brother Paul who patiently followed my research in this field and provided me with his valuable comments and corrections, and, finally, to my wife Tanya, whose advice and assistance in editing my books were invaluable. Many great memories about my wonderful late parents and very wise grandparents provided me with a needed moral support in writing my books.

ACKNOWLEDGEMENTS

I am grateful to several scientists for making their valuable and stimulating comments on my idea of multi-level Universe and also on my spiral spacetime model. Among them are:

Professor Carlo Rovelli (Department of Physics and Astronomy of the University of Pittsburgh),
Professor Gregory M. Townsend (Department of Physics of the University of Akron),
Professor Clifford Taubes (Department of Mathematics, Harvard University)
Professor Rudolph Hwa (Department of Physics, University of Oregon),
Dr. Blair M. Smith (Innovative Nuclear Space Power and Propulsion Institute, University of Florida),
Professor Edward F. Redish (Department of Physics, University of Maryland),
Professor David F. Measday (Department of Physics and Astronomy, University of British Columbia, Canada),
Dr. Eric Carlson (Department of Physics & Astronomy, Wake Forest University),
Professor Warren Siegel (Department of Physics and Astronomy, Stony Brook University).
Dr. Charles E. Hyde-Wright, Professor of Physics, Old Dominion University,
Dr. John G. Fetkovich, Professor Emeritus (Department of Physics, Carnegie Mellon University),
Dr. Yan Corlett, York University, Canada,
Dr. Igor Sokolov, Assistant Professor of Department of Physics, Clarkson University and
Dr. Thomas Love, California State University at Dominguez Hills.

Thanks to the recommendations made by Dr. Akhlesh Lakhtakia (Department of Engineering and Mechanics, the Pennsylvania State University), I was able to publish the first three papers describing the earliest versions of my theory in *Speculations in Science and Technology* in 1996-1998. I appreciate very much my interesting discussions with late Marvin Solit, Director of Foundations for New Directions, and his ideas about possible space orientations of toryces. I will always remember very stimulating discussions with the late Distinguished Professor Eli Gorelik of the University of Pittsburgh, Pennsylvania about a crystal structure of a nucleon core.

I value greatly my two meetings with Dr. Arlie Oswald Petters, the Benjamin Powell Professor and Professor of Mathematics, Physics and Business Administration at Duke University held at the Duke University in October 2015. After an open-minded review of a basic concept of my theory, Dr. Petters offered me several practical recommendations on how to introduce it to the academic world. Finally, let me express my gratitude to Dr. John Kern II, Chair, Dr. Anna Haensch and Ms. Larisa Shtrahman, Instructor of Mathematics (all from Department of Mathematics & Computer Science of Duquesne University, Pittsburgh, PA), for providing me with an opportunity to conduct a seminar on mathematical aspects of my theory in their department on February 15, 2017. I also appreciate an opportunity to present my theory at the Materials Science & Technology (MS&T) 2017 Conference held in Pittsburgh, PA and at the next Materials Science & Technology (MS&T) 2018 Conference to be held in Columbus OH.

Vladimir B. Ginzburg

This book is dedicated to Walter Russell, the American self-educated scientist, architect and artist who envisioned the spacetime origin of the Universe in his book:

That the man calls matter, or substance, has no existence whatsoever. So-called matter is but waves of the motion of light, electrically divided into opposed pairs, then electrically conditioned and patterned into what we call various substances of matter. Briefly put, matter is the motion of light, and motion is not substance. It only appears to be. Take motion away and there would not be even the appearance of substance.

Walter Russell

The author of *The Secret of Light*.

INTRODUCTION - *TORYX & HELYX IN-BRIEF*

This book describes the unique properties of two 4D spiral spacetimes called *toryx* and *helyx.* The mathematical description of the toryx and helyx is at the core of the Unified Spacetime Multiverse (USM) theory that the author of this book had been developing since 1993. Parts 1 and 3 of this book describe the abstract mathematics of toryx and helyx. Parts 2 and 4 provide several examples of modeling the structures and properties of entities of both micro- and macro-worlds of our own Universe and other spacetime levels of the Multiverse.

The efficiency of any mathematical model of a natural phenomenon is defined by its ability to calculate accurately as many as possible properties of the modeled phenomena based on as a few as possible assumptions. A high efficiency of the USM theory is achieved thanks to the *unique polarization, quantum, unification and generic properties* of the proposed toryx and helyx allowing us to model structures and properties of entities of both micro- and macro-worlds of various levels of the Multiverse by using mostly the same spacetime equations, explaining why the USM theory is called "unified."

1. Polarization Properties of Toryx & Helyx

As shown in Fig. 1, the toryx is the 4D spiral spacetime made up of a double-circular *leading string* with the radius r_1 propagating with the velocity V_1 and a double-toroidal *trailing string* with the radius r_2 located inside a *spherical boundary* and propagating synchronously with the circular leading string. The spiral velocity of trailing string V_2 is equal to the velocity of light c. It consists of two components: the rotational velocity V_{2r} and the translational velocity V_{2t} that is equal to the velocity of leading string V_1. Also shown on Fig. 1 are the wavelength of trailing string λ_2, the steepness angle of trailing string φ_2 and the toryx eye radius r_0.

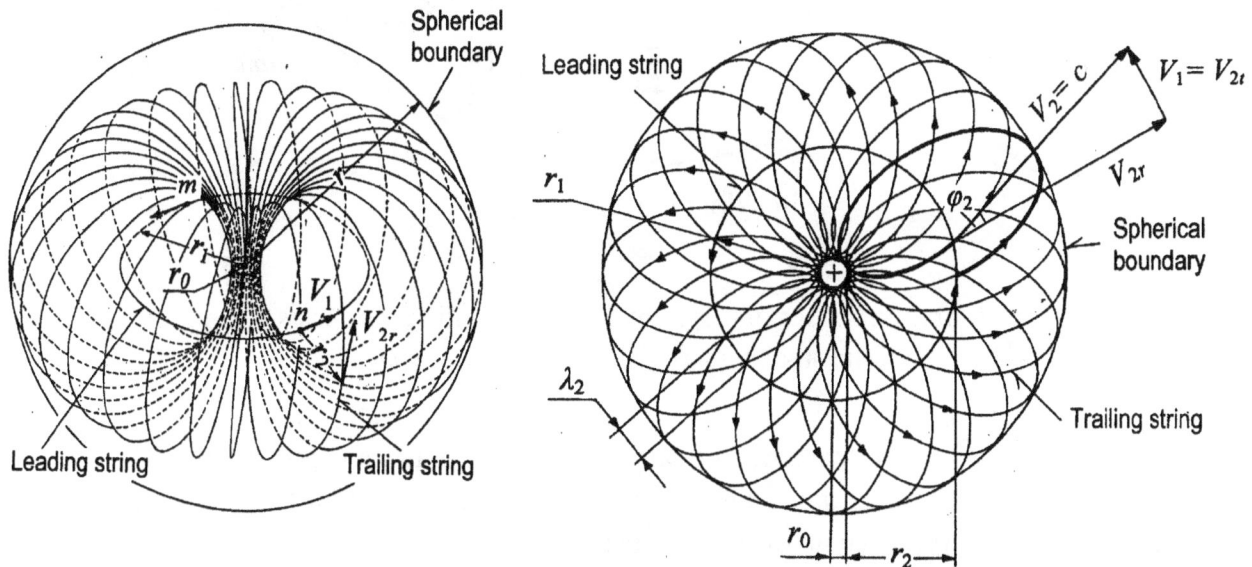

Figure 1. Toryx views: isometric (left) and top (right).

The toryx spacetime postulates provide the toryx with two fundamental capabilities. Firstly, they limit the toryx degrees of freedom, while assuring the simplest way of establishing the relationships between the toryx main parameters. Secondly, they provide the toryx with a unique capability to be transformed into four topologically-polarized spacetime forms as the radius r_1 of its leading string decreases from positive to negative infinity as shown in Fig. 2.

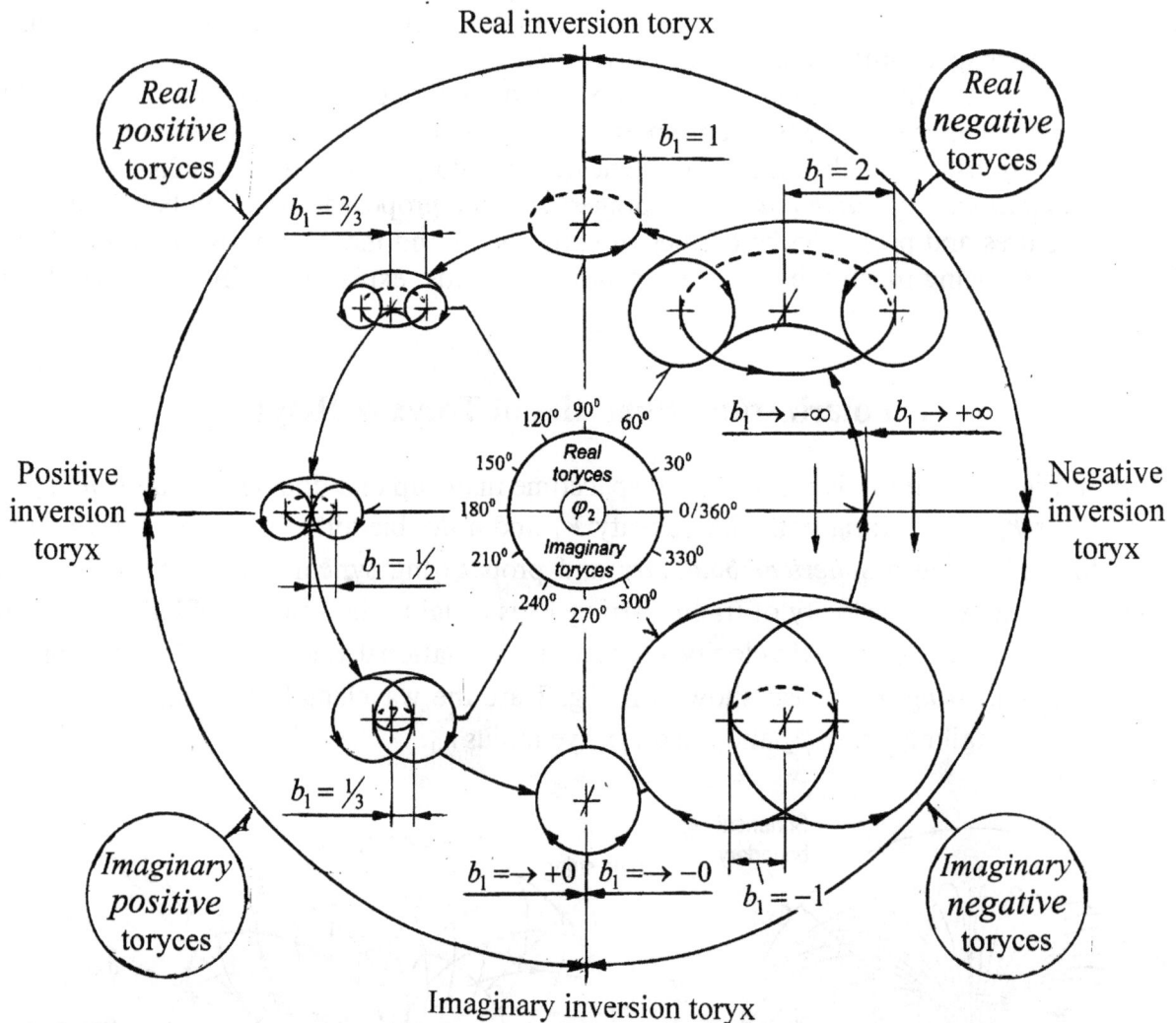

Figure 2. Metamorphoses of toryx leading and trailing strings as a function of the steepness angle of trailing string φ_2.

Analysis of derived toryx spacetime equations reveals the unique polarization properties of toryces. Figure 2 shows that as the radius of the toryx leading string decreases from positive to negative infinity, the steepness angle of trailing string φ_2 increases from 0^0 to 360^0. Consequently, four kinds of topologically-polarized toryces are formed:

- Real negative toryces - ***Trailing string*** becomes inverted at $\varphi_2 = 90^0$
- Real positive toryces - ***Wavelength of trailing string*** becomes inverted at $\varphi_2 = 180^0$
- Imaginary positive toryces - ***Leading string*** becomes inverted at $\varphi_2 = 270^0$
- Imaginary negative toryces - ***Entire toryx*** becoming inverted at $\varphi_2 = 360^0$.

As shown in Fig. 3 the helyx basic structure is made up of a ***leading string*** O_1O_1 and a double-helical ***trailing string***, both residing inside a ***cylindrical boundary***. The leading string propagates with the translational velocity $\tilde{V}_{1t}$ along a line O_1O_1 with the radius $\tilde{r}_1 \to \infty$. The double-helical trailing string with the radius $\tilde{r}_2$ rotates around the leading string with the rotational velocity $\tilde{V}_{2r}$, while propagating along the line O_1O_1 with the translational velocity $\tilde{V}_{2t}$. Its spiral velocity $\tilde{V}_2$ is equal to the velocity of light c. Also shown in Fig. 3 are the wavelength of trailing string $\tilde{\lambda}_2$, the apex angle of trailing string $\tilde{\varphi}_2$ and the helyx eye radius $\tilde{r}_0$.

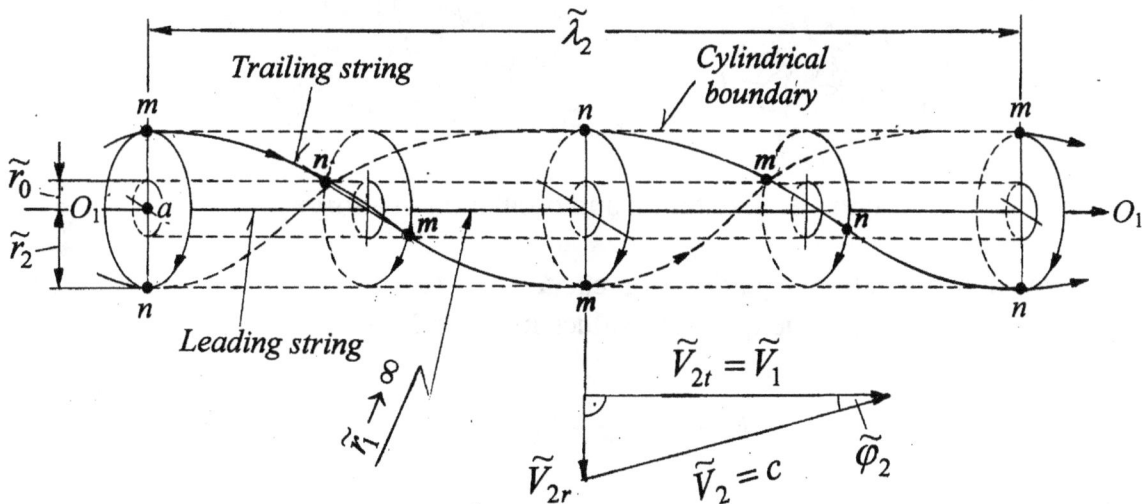

Figure 3. Helyx basic structure.

As the spiral length of helyx leading string decreases from positive to negative infinity, the apex angle of helyx trailing string $\tilde{\varphi}_2$ increases from 0 to 360 degrees.

Consequently, four kinds of topologically-polarized helyces are formed:

- Real negative helyces - ***Wavelength of trailing string*** becomes inverted at $\tilde{\varphi}_2 = 90^0$
- Real positive helyces - ***Radius of trailing string*** becomes inverted at $\tilde{\varphi}_2 = 180^0$
- Imaginary positive helyces - ***Spiral length of trailing string*** becomes inverted at $\tilde{\varphi}_2 = 270^0$
- Imaginary negative helyces - ***Entire helyx*** becomes inverted when $\tilde{\varphi}_2 = 0^0 / 360^0$.

2. Quantum Properties of Toryces

Toryces change their dimensions in quantum steps by excitation and oscillation. During excitation of a toryx the radius of its leading string r_1 increases, while its eye radius r_0 remains constant. During oscillation of a toryx, both its radius of leading string r_1 and its eye radius r_0 change proportionally as shown in Fig. 4.

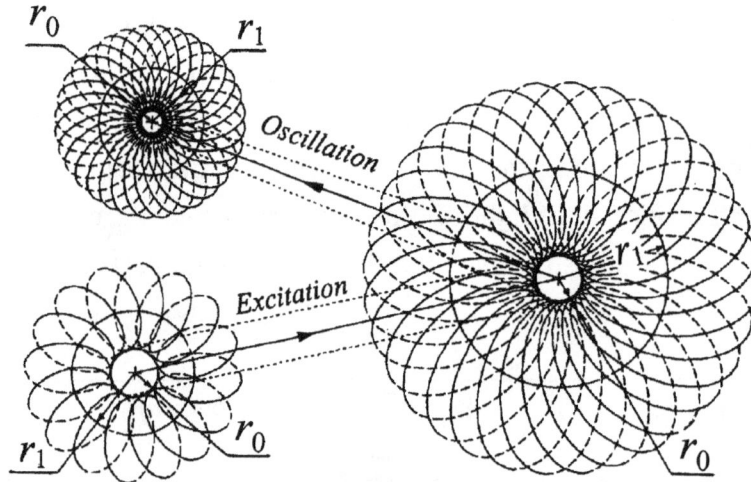

Figure 4. Excitation and oscillation of a toryx.

The derived quantization equations of excited toryces are based on the following basic quantization equation defining the quantum values for the relative radius of leading string b_1 of a real negative toryx:

$$b_1 = \frac{r_1}{r_0} = z = 2(n\Lambda)^m \qquad (1)$$

where
   z = roryx quantization parameters
   $m \to 0, 1, 2, \ldots$ toryx quantum states depending on the spacetime levels of the Multiverse
   $n \to 0, 1, 2, \ldots$ toryx linear excitation quantum states
   $\Lambda = 137\ldots$ ***toryx quantization constant***.

The derived quantization equations are based on three proposed limitations of degrees of freedom of real negative ***lambda, harmonic*** and ***golden toryces*** shown in Table 1.

Table 1. Limitations of degrees of freedom of excited toryces.

Lambda toryx	Harmonic toryx	Golden toryx
$b_1 = z = 2(n\Lambda)^m$	$b_1 = z = 2 + n$	$b_1 = z = 2 + n/\phi$

3. Unification Properties of Toryces & Helyces

The elementary matter particles (trons) are formed by the unification of topologically-polarized toryces. Four trons are formed by the unification of adjacent toryces: *electrons, positrons, ethertrons* and *singulatrons* as shown in Fig. 5.

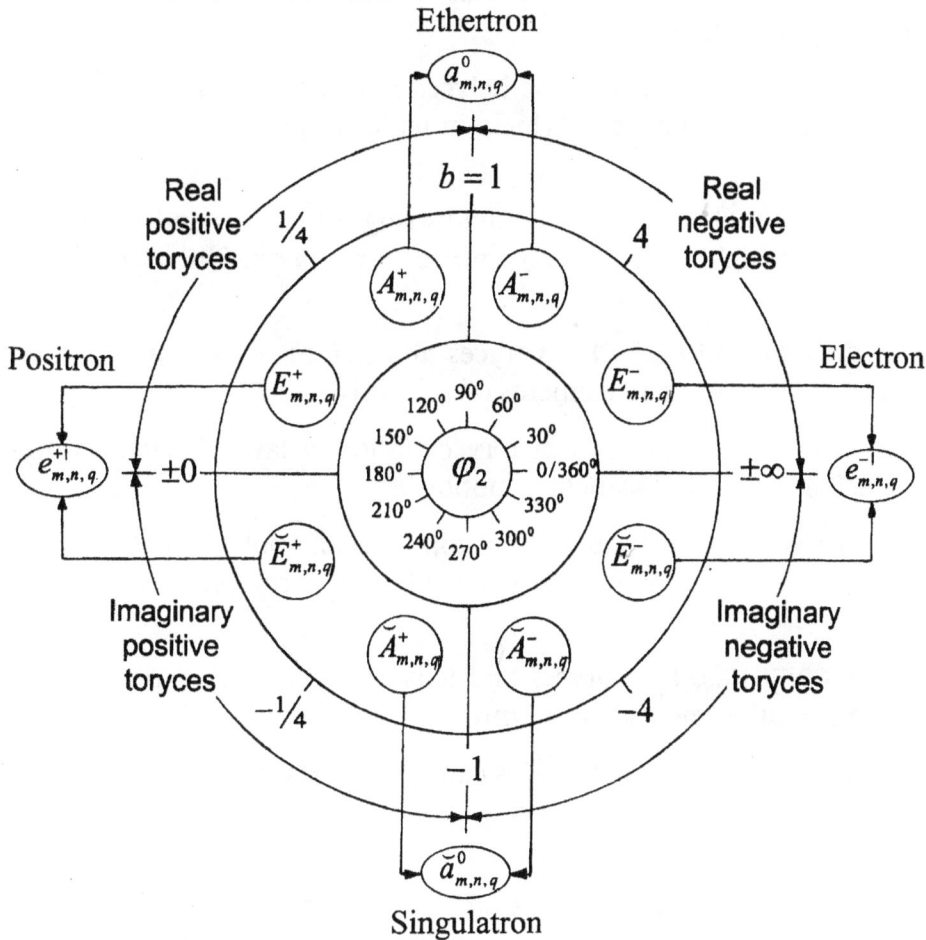

Figure 5. Formation of topologically-polarized trons.

Topologically-polarized helyces are formed when their parental topologically-polarized toryces are transferred from higher to lower excitation and oscillation quantum states. The elementary radiation particles (tons) are formed by the unification of topologically-polarized helyces.

The elementary matter and radiation particles sustain their existence thanks to the absorption and release of spacetimes by their respective constituent polarized toryces and helyces.

4. Generic Properties of Toryces & Helyces

Spacetime properties – Properties of toryces and helyces can be completely described by using the space and time parameters. The spacetime properties can be found in all elements of our Universe.

Motion – Toryces and helyces are in a state of continuous motion. All elements in our Universe are in the motion state.

Spirality – Trailing strings of toryces and helyces propagate along spiral paths. Each celestial body spirals around another celestial body which, in its turn, spirals around another celestial body, etc.

Propagation with velocity of light – Trailing strings of toryces and helyces propagate along spiral paths with velocity of light. All electromagnetic waves in our Universe propagate in vacuum with velocity of light.

Limitation of spacetime freedom – Toryces and helyces have limited spacetime freedom. The spacetime freedom is limited in all elements of our Universe.

Planetary motion – Leading strings of toryces follow a law of planetary motion. Atomic electrons and planets follow a law of planetary motion.

Polarization – Toryces and helyces exist in topologically-polarized states. Atoms are made up of electrically-polarized electrons and nuclei, while many complex elements are made up of chemically-polarized acids and bases.

Expansion and contraction – Toryces and helyces are able to expand and contract. This capability can be found in all elements of our Universe.

Stable coexistence – Toryces and helyces can stably coexist with their respective oppositely-polarized toryces. Electrons and protons stably coexist in atoms, while acids and bases stably coexist in DNA.

Self-preservation – Toryces and helyces are able to self-preserve their existence. All stable elements of our Universe exist because they are able to self-preserve themselves.

Absorption and release of spacetime – Toryces and helyces are able to absorb and release spacetime to sustain their existence. All elements of our Universe sustain their existence by absorption and release of spacetime, or energy in physical terms.

Quantization – Toryces and helyces exist in quantum spacetime states. This property extends to the atomic electrons.

Emission of radiation – Excited toryces are capable of emitting the radiation in the form of helyces. Excited atomic electrons emit electromagnetic radiation in the form of photons.

The polarization, quantum, unification and generic spacetime properties of toryces and helyces make them fundamentally different from the prime elements of nature proposed so far.
Take these properties away and the Universe, as we know it, will cease to exist.

5. Tasks Accomplished by the USM Theory

Thanks to the unique topological, quantum and generic properties of toryx and helyx, the USM theory is able to accomplish the following tasks:

- To identify the conditions for the creation of topologically-polarized toryces from quantum vacuum
- To model the formation of self-sustained harmonic elementary matter particles
- To model the formation of mutually-sustained elementary matter particles by the unification of polarized toryces
- To model quantum states of toryces
- To model the formation of nucleons and atomic electrons
- To model the emission of elementary radiation particles by the excited and oscillated elementary matter particles
- To model the formation of stable elementary particles by a balanced absorption and release of spacetime
- To demonstrate that our ordinary Universe is merely one of the quantum levels of the Multiverse
- To identify the levels of the Multiverse at which the elementary radiation particles can propagate with velocities significantly exceeding the velocity of light
- To model structures and properties of entities of both micro- and macro-worlds of various spacetime levels of the Multiverse by using mostly the same spacetime equations.
- To reduce significantly the complexity of employed mathematics by using the 3-dimensional (3D) elementary mathematics taught in high schools with some modifications making it applicable to the 4D mathematics of the toryx and helyx.
- To express the properties of entities of the Multiverse with both the *subjective physical units* invented by human beings and with the *objective spacetime units* that can readily be discovered by the inhabitants of all advanced civilizations in the Multiverse
- To prove a long-standing proposition that what current theories of physics describe as matter, field, charge, mass, electromagnetic radiation and gravity are merely metamorphoses of spiral spacetimes.

6. Several Possible Areas of Applications of the USM Theory

Abstract mathematics – The abstract mathematics of the USM theory may stimulate the development of a two new branches of mathematics that I called the ***spiral spacetime topology*** and the ***spiral spacetime number theory***:

- Spiral spacetime topology is a study of topological transformations of 4D spiral spacetimes toryces and helyces, including their topological inversions.
- Spiral spacetime number theory is a supplementary tool for the spiral spacetime topology. Applied to both real and imaginary numbers, the theory describes symmetrical relationships between spacetime parameters of both toryces and helyces with a use of circular number lines in which the unity (± 1) extends equally towards both infinity ($\pm\infty$) and infinility (± 0).

Applied mathematics – The predictions of applied mathematics of the USM theory may stimulate physicists and astrophysicists to model alternatively many secrets of nature and to accomplish the tasks similar to those described in Section 4.

Experimental particle physics and astrophysics – The predictions of the USM theory can be used to encourage the experimental particle physicists and astrophysicists to re-examine the results of their experiments produced by the most advanced technology, like the Hubble telescope and the CERN proton-proton collider.

Material science and technology – The predictions of the USM theory can be used to stimulate the discovery of materials that belong to other spacetime levels of the Universe, besides the ordinary matter level, with properties significantly better suitable for their applications in the technology of the forthcoming spacetime age.

Communication science & technology – The predictions of the USM theory can be also used to stimulate the development of communication systems utilizing the materials that belong to other spacetime levels of the Multiverse, besides the ordinary matter level, and capable of emitting and receiving the information with velocities significantly exceeding the velocity of light.

Energy science & technology – The predictions of the USM theory can be also used to stimulate the engineering of much more efficient methods of generation and storage of energy with a minimal detrimental impact on environment.

The rest is commentary, as described in this book.

PART 1

Abstract
Mathematics
of a Toryx

1. TORYX BASIC STRUCTURE & PARAMETERS

CONTENTS

1.1 Toryx Basic Structure

Toryx basic structure consists of a double-circular *leading string* and a double-toroidal *trailing string*, both residing inside a *spherical boundary* as shown in Figs. 1.1.1 – 1.1.3.

Toryx leading string – The toryx double-circular leading string appears like two circular traces with the radius r_1 left by the moving points m and n shown in Figure 1.1.1. It can be thought as a particular case of a double-helical spiral in which the translational velocity V_{1t} is equal to zero, so the rotational velocity V_{1r} is equal to the spiral velocity V_1. Thus,

$$V_{1t} = 0 \qquad\qquad (1.1\text{-}1)$$

$$V_{1r} = V_1 \qquad\qquad (1.1\text{-}2)$$

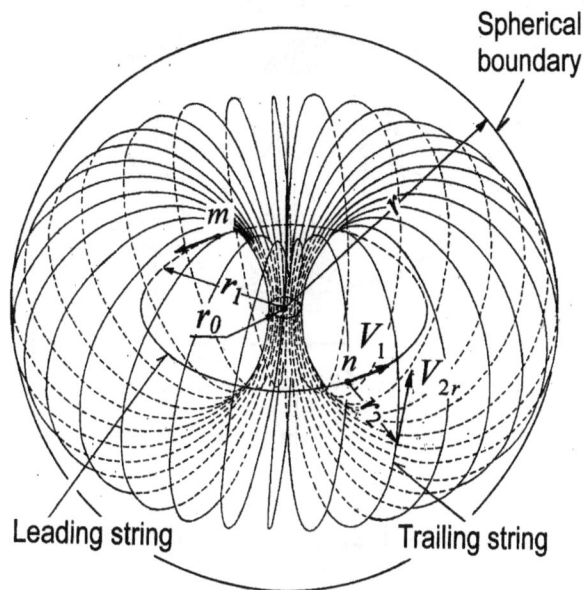

Figure 1.1.1. Isometric view of a toryx.

Toryx trailing string – The toryx double-toroidal trailing string has the radius r_2 and a circular opening at the toryx center with the radius r_0 called the ***toryx eye***. It can be thought as a dynamic double-toroidal spiral in which each branch propagates along its toroidal spiral path with the spiral velocity V_2 that has two components, the translational velocity V_{2t} and the rotational velocity V_{2r}, with all three velocities related to each other by the Pythagorean Theorem:

$$V_2 = \sqrt{V_{2t}^2 + V_{2r}^2} \tag{1.1-3}$$

The trailing string propagates synchronously with the leading string. Therefore, the translational velocity of trailing string V_{2t} is equal to the rotational velocity V_{1r} and the spiral velocity V_1 of leading string as given by:

$$V_{2t} = V_{1r} = V_1 \tag{1.1-4}$$

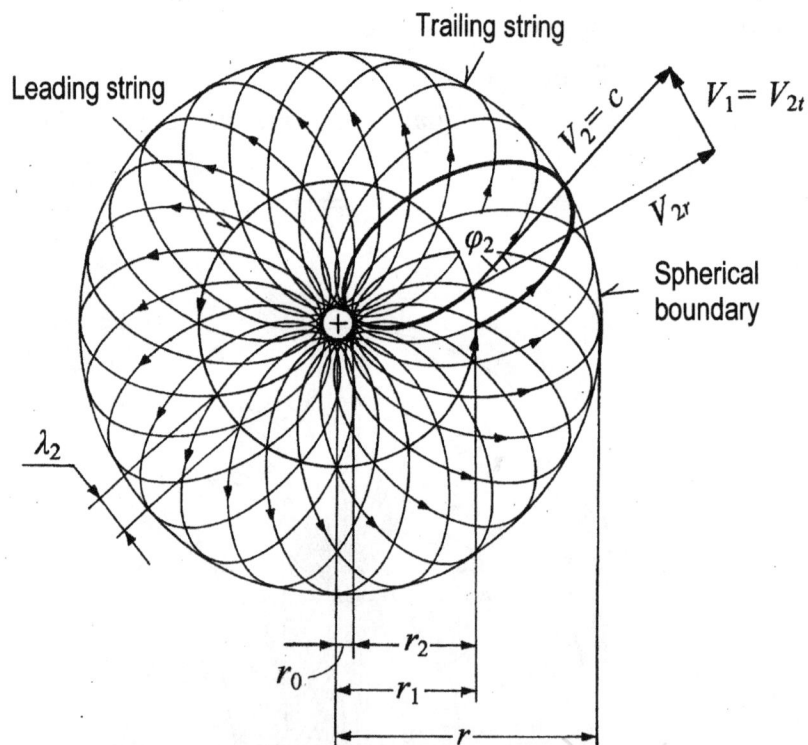

Figure 1.1.2. Top view of a toryx.

In Fig. 1.1.2, φ_2 is the ***steepness angle*** of toryx trailing string corresponding to the middle point *a* of toryx trailing string.

The radius of spherical boundary r is equal to:

$$r = r_1 + r_2 \qquad (1.1\text{-}5)$$

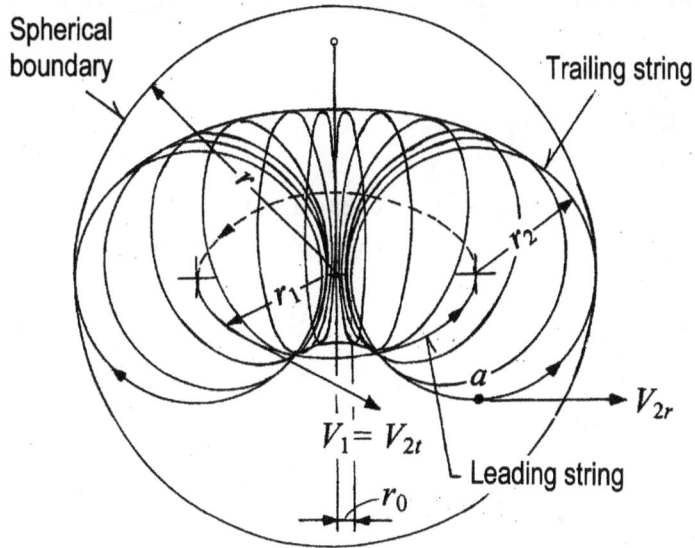

Figure 1.1.3. Cross-section of a toryx.

Toryx spins – Toryx has two spins, leading and trailing, with both of them defined by the right-hand rule as shown in Fig. 1.1.4. The toryx leading and trailing spins depend on the directions of the rotational velocities of toryx leading string V_{1r} and toryx trailing string V_{2r}.

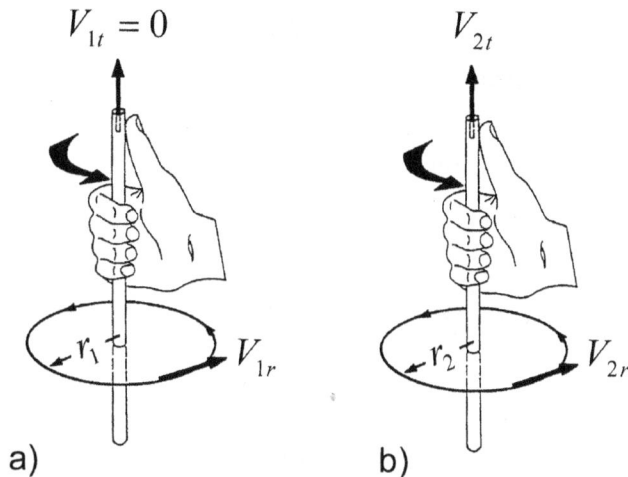

Figure 1.1.4. The toryx leading spin (a) and trailing spin (b) defined by the right-hand rule.

1.2 Toryx Spacetime Parameters in Absolute Units

The values of some toryx spacetime parameters are dependent on the position of a point of measurement of these parameters along the toryx trailing string. In the nomenclature of toryx parameters shown below, the symbols of the position-dependent parameters marked with star (*) correspond to the middle point a of toryx trailing string shown in Fig. 1.1.3.

f_0 = toryx base frequency
f_1 = frequency of toryx leading string
f_2 = frequency of toryx trailing string
L_1 = spiral length of one winding of toryx leading string
L_2 = spiral length of one winding of toryx trailing string
L_0 = circular length of toryx eye
r = radius of toryx spherical boundary
r_0 = toryx eye radius
r_1 = radius of toryx leading string
r_2 = radius of toryx trailing string
T_1 = period of toryx leading string
T_2 = period of toryx trailing string
V_1 = spiral velocity of toryx leading string
V_{1t} = translational velocity of toryx leading string
V_{1r} = rotational velocity of toryx leading string
V_2 = spiral velocity of toryx trailing string
V_{2t} = translational velocity of toryx trailing string*
V_{2r} = rotational velocity of toryx trailing string*
w_1 = the number of windings of toryx leading string
w_2 = the number of windings of toryx trailing string
λ_1 = wavelength of toryx leading string
λ_2 = wavelength of toryx trailing string*
φ_1 = steepness angle of toryx leading string
φ_2 = steepness angle of toryx trailing string*.

1.3 Toryx Spacetime Postulates in Absolute Units

The toryx spacetime postulates include three fundamental equations limiting the degrees of freedom of several toryx parameters (see Exhibit 1.3). These postulates provide the simplest way to derive the relationships between all spacetime parameters of toryx. In spite of their outmost simplicity, these postulates provide toryces with amazing spacetime properties, including a capability to exist in four unique topologically-polarized states within the range of the radius of toryx leading strings r_1 extending from negative to positive infinity $(-\infty < r_1 < +\infty)$ as will be described in Chapter 5.

Exhibit 1.3. Toryx spacetime postulates in absolute units.

- The length of one winding of toryx trailing string L_2 is equal to the length of one winding of toryx leading string L_1:

$$L_2 = L_1 = 2\pi r_1 \qquad (1.3\text{-}1)$$

- The toryx eye radius r_0 is constant:

$$r_0 = r_1 - r_2 = const. \qquad (1.3\text{-}2)$$

- The spiral velocity of toryx trailing string V_2 is constant at each point of its spiral path:

$$V_2 = \sqrt{V_{2t}^2 + V_{2r}^2} = c = const. \qquad (1.3\text{-}3)$$

Figure 1.3 shows the toryx spacetime parameters in absolute units corresponding to the middle point a of toryx trailing string (Fig. 1.1.3).

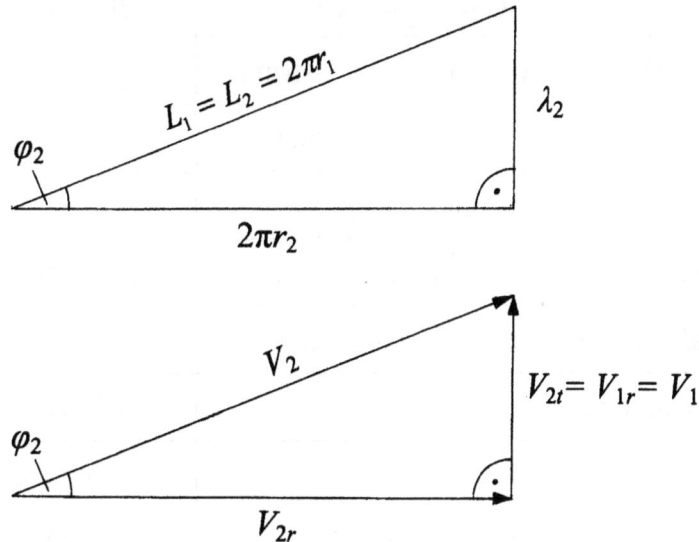

Figure 1.3. Toryx spacetime parameters in absolute units corresponding to the middle point a of toryx trailing string (Fig. 1.1.3).

1.4 Toryx Spacetime Parameters in Relative Units

The toryx spacetime parameters can be simplified by expressing them in relative units in respect to the constant toryx parameters: the toryx eye radius r_0, the velocity of light c and the toryx base frequency f_0 as shown in Table 1.4.

Table 1.4. Toryx relative spacetime parameters.

Toryx relative parameters	Equations
Radius of toryx spherical boundary	$b = r/r_0$ (1.4-1)
Radius of toryx leading string	$b_1 = r_1/r_0$ (1.4-2)
Radius of toryx trailing string	$b_2 = r_2/r_0$ (1.4-3)
Length of toryx leading string	$l_1 = L_1/2\pi r_0$ (1.4-4)
Length of toryx trailing string	$l_2 = L_2/2\pi r_0$ (1.4-5)
Period of one winding of toryx leading string	$t_1 = T_1 f_0$ (1.4-6)
Period of one winding toryx trailing string	$t_2 = T_2 f_0$ (1.4-7)
Spiral velocity of toryx leading string	$\beta_1 = V_1/c$ (1.4-8)
Translational velocity of toryx leading string	$\beta_{1t} = V_{1t}/c$ (1.4-9)
Rotational velocity of toryx leading string	$\beta_{1r} = V_{1r}/c$ (1.4-10)
Spiral velocity of toryx trailing string	$\beta_2 = V_2/c$ (1.4-11)
Translational velocity of toryx trailing string	$\beta_{2t} = V_{2t}/c$ (1.4-12)
Rotational velocity of toryx trailing string	$\beta_{2r} = V_{2r}/c$ (1.4-13)
Frequency of toryx leading string	$\delta_1 = f_1/f_0$ (1.4-14)
Frequency of toryx trailing string	$\delta_2 = f_2/f_0$ (1.4-15)
Wavelength of toryx leading string	$\eta_1 = \lambda_1/2\pi r_0$ (1.4-16)
Wavelength of toryx trailing string	$\eta_2 = \lambda_2/2\pi r_0$ (1.4-17)

The **toryx base frequency** f_0 corresponds to the case when $r_1 = r_0$ and it is equal to:

$$f_0 = \frac{c}{2\pi r_0}$$

(1.4-18)

1.5 Toryx Spacetime Postulates in Relative Units

Exhibit 1.5 shows three toryx spacetime postulates in relative units.

Exhibit 1.5. Toryx spacetime postulates in relative units.

- The relative length of one winding of toryx trailing string l_2 is equal to the relative length of one winding of toryx leading string l_1:

$$l_2 = l_1 \qquad (1.5\text{-}1)$$

- The toryx relative eye radius b_0 is equal to 1:

$$b_0 = b_1 - b_2 = 1 \qquad (1.5\text{-}2)$$

- The relative spiral velocity of toryx trailing string β_2 is equal to 1 at each point of its spiral path:

$$\beta_2 = \sqrt{\beta_{2t}^2 + \beta_{2r}^2} = 1 \qquad (1.5\text{-}3)$$

Figure 1.5 shows the toryx relative spacetime parameters corresponding to the middle point a of toryx trailing string (Fig. 1.1.3).

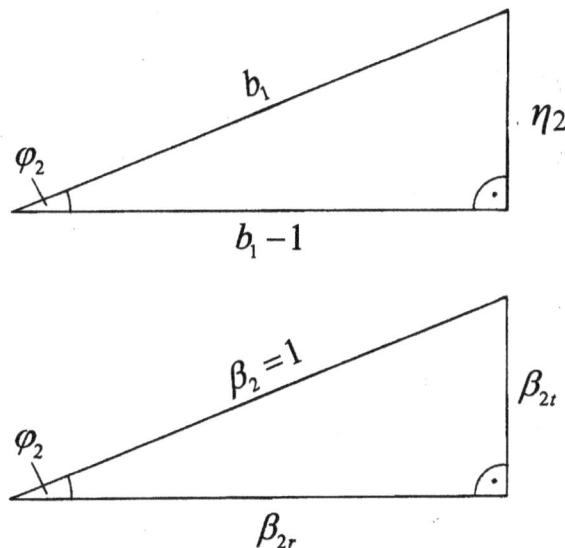

Figure 1.5. Toryx relative spacetime parameters corresponding to the middle point a of toryx trailing string (Fig. 1.1.3).

Chapter 1

1.6 Summary of Derived Toryx Equations

Table 1.6.1 provides a summary of derived equations for main relative spacetime parameters of toryx as functions of the relative radius of toryx leading string b_1.

Table 1.6.1. Equations for relative spacetime parameters of toryx leading and trailing strings as functions of the relative radius of toryx leading string b_1.

Relative parameter	Leading string Eq. (a)	Trailing string Eq. (b)
Radius Eq. (1.6-1)	$b_1 = \dfrac{r_1}{r_0}$	$b_2 = \dfrac{r_2}{r_0} = b_1 - 1$
Wavelength Eq. (1.6-2)	$\eta_1 = \dfrac{\lambda_1}{2\pi r_0} = b_1$	$\eta_2 = \dfrac{\lambda_2}{2\pi r_0} = \sqrt{2b_1 - 1}$
Length of one winding Eq. (1.6-3)	$l_1 = \dfrac{L_1}{2\pi r_0} = b_1$	$l_2 = \dfrac{L_2}{2\pi r_0} = b_1$
Steepness angle Eq. (1.6-4)	$\varphi_1 = 0$	$\cos s\varphi_2 = \dfrac{b_1 - 1}{b_1}$
The number of windings Eq. (1.6-5)	$w_1 = 1$	$w_2 = \dfrac{b_1}{\sqrt{2b_1 - 1}}$
Translational velocity Eq. (1.6-6)	$\beta_{1t} = \dfrac{V_{1t}}{c} = 0$	$\beta_{2t} = \dfrac{V_{2t}}{c} = \dfrac{\sqrt{2b_1 - 1}}{b_1}$
Rotational velocity Eq. (1.6-7)	$\beta_{1r} = \dfrac{V_{1r}}{c} = \dfrac{\sqrt{2b_1 - 1}}{b_1}$	$\beta_{2r} = \dfrac{V_{2r}}{c} = \dfrac{b_1 - 1}{b_1}$
Spiral velocity Eq. (1.6-8)	$\beta_1 = \dfrac{V_1}{c} = \dfrac{\sqrt{2b_1 - 1}}{b_1}$	$\beta_2 = \dfrac{V_2}{c} = 1$
Frequency Eq. (1.6-9)	$\delta_1 = \dfrac{f_1}{f_0} = \dfrac{\sqrt{2b_1 - 1}}{b_1^2}$	$\delta_2 = \dfrac{f_2}{f_0} = \dfrac{1}{b_1}$
Period Eq. (1.6-10)	$t_1 = T_1 f_0 = \dfrac{b_1^2}{\sqrt{2b_1 - 1}}$	$t_2 = T_2 f_0 = b_1$

The values of η_2, β_{2t} and β_{2r} correspond to the middle point a of toryx trailing string (Fig. 1.1.3). In Eq. (1.6-4b), $\cos s\varphi_2$ is the toryx trigonometric function. It relates to the trigonometric function $\cos\varphi_2$ of elementary mathematics as follows:

$$\cos s\varphi_2 = \cos\varphi_2 \quad (0 < \varphi_2 < 180^0) \tag{1.6-11}$$

$$\cos s\varphi_2 = 1/\cos\varphi_2 \quad (180^0 < \varphi_2 < 360^0) \tag{1.6-12}$$

The relative radius of toryx spherical boundary b is equal to:

$$b = \frac{r}{r_0} = 2b_1 - 1 \tag{1.6-13}$$

Table 1.6.2 provides a summary of derived equations for relative spacetime parameters of toryx leading and trailing strings as a function of the steepness angle of toryx trailing string φ_2.

Table 1.6.2. Equations for relative spacetime parameters of toryx leading and trailing strings as functions of the steepness angle of toryx trailing string φ_2.

Relative parameter	Leading string Eq. (a)	Trailing string Eq. (b)
Radius Eq. (1.6-14)	$b_1 = \dfrac{1}{1 - \cos s\varphi_2}$	$b_2 = \dfrac{\cos s\varphi_2}{1 - \cos s\varphi_2}$
Wavelength Eq. (1.6-15)	$\eta_1 = 0$	$\eta_2 = \sqrt{\dfrac{1 + \cos s\varphi_2}{1 - \cos s\varphi_2}}$
Length of one winding (Eq. 1.6-16)	$l_1 = \dfrac{1}{1 - \cos s\varphi_2}$	$l_2 = \dfrac{1}{1 - \cos s\varphi_2}$
The number of windings Eq. (1.6-17)	$w_1 = 1$	$w_2 = \dfrac{1}{\sin s\varphi_2}$
Translational velocity Eq. (1.6-18)	$\beta_{1t} = \dfrac{V_{1t}}{c} = 0$	$\beta_{2t} = \sin s\varphi_2$
Rotational velocity Eq. (1.6-19)	$\beta_{1r} = \sin s\varphi_2$	$\beta_{2r} = \cos s\varphi_2$
Spiral velocity Eq. (1.6-20)	$\beta_1 = \sin s\varphi_2$	$\beta_2 = 1$
Frequency Eq. (1.6-21)	$\delta_1 = \sin s\varphi_2(1 - \cos s\varphi_2)$	$\delta_2 = 1 - \cos s\varphi_2$
Period Eq. (1.6-22)	$t_1 = \dfrac{1}{\sin s\varphi_2(1 - \cos s\varphi_2)}$	$t_2 = \dfrac{1}{1 - \cos s\varphi_2}$

The relative radius of toryx spherical boundary b is equal to:

$$b = \frac{1 + \cos s\varphi_2}{1 - \cos s\varphi_2}$$

(1.6-23)

1.7 Toryx Law of Planetary Motion

The first strong indication that derived toryx spacetime equations were relevant to a real world comes from Eq. (1.6-8a). This equation establishes a relationship between the relative velocity of the toryx leading string β_1 and the relative radius of this string b_1 that is rewritten below:

$$\beta_1 = \frac{V_1}{c} = \frac{\sqrt{2b_1 - 1}}{b_1}$$

(1.7-1)

For the case when $b_1 \gg 1$, the above equation reduces to the form:

$$\beta_1 = \frac{V_1}{c} = \sqrt{\frac{2}{b_1}}$$

(1.7-2)

Let us show that Eq. (1.7-2) expresses the Third Kepler's law of planetary motion.

The velocity of the toryx leading string V_1 is equal to:

$$V_1 = \frac{2\pi r_1}{T_1}$$

(1.7-3)

From Eqs. (1.6-1a), (1.7-2) and (1.7-3):

$$\frac{2\pi r_1}{T_1 c} = \sqrt{\frac{2r_0}{r_1}}$$

(1.7-4)

After squaring both parts of Eq. (1.7-4):

$$\frac{4\pi^2 r_1^2}{T_1^2 c^2} = \frac{2r_0}{r_1}$$

we obtain:

$$r_1^3 = \frac{r_0 c^2}{2\pi^2} T_1^2 \tag{1.7-5}$$

Let k to be equal to:

$$k = \frac{r_0 c^2}{2\pi^2} \tag{1.7-6}$$

From Eqs. (1.3-2) and (1.3-3), both r_0 and c are constant. Therefore k is constant too. Consequently, Eqs. (1.7-5) and (1.7-6) yield Kepler's third law of planetary motion:

$$r_1^3 = kT_1^2 \tag{1.7-7}$$

Both Eqs. (1.7-7) and (1.7-2) express Kepler's law of planetary motion with only one difference: In Eq. (1.7-7) parameters are expressed in absolute units, while in Eq. (1.7-2) in relative units. Thus, Kepler's third law of planetary motion expressed by Eq. (1.7-2) is applied for large relative orbital radii ($b_1 \gg 1$) and can be treated as a particular case of the ***toryx law of planetary motion*** that is applied for the orbital radii extending from negative to positive infinity.

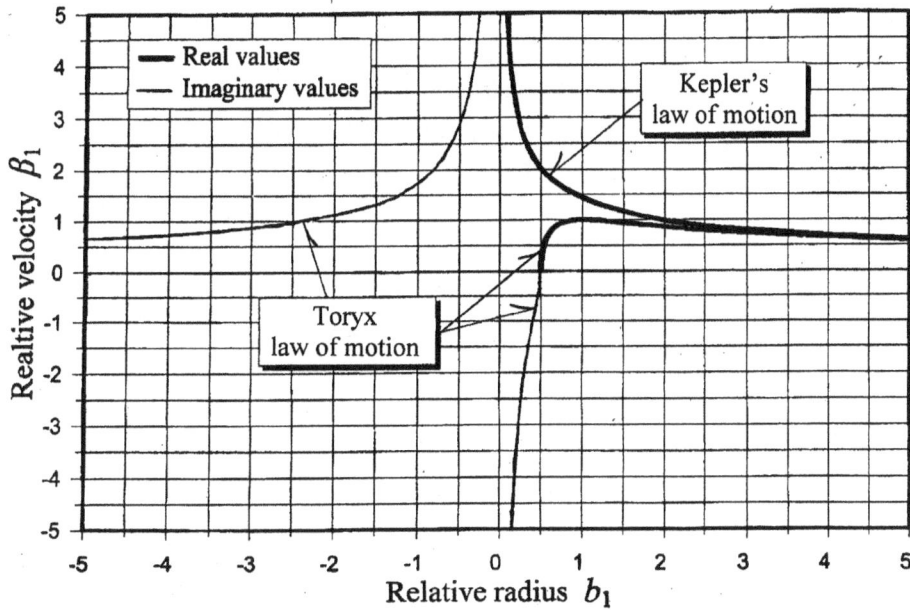

Figure 1.7. Toryx law of planetary motion versus Kepler's third law of planetary motion.

Figure 1.7 shows plots of Eqs. (1.7-1) and (1.7-2) expressing respectively the toryx law of planetary motion and Kepler's third law of planetary motion. The highlights of these plots are:

- The toryx law of motion is described by Eq. (1.7-1). It is applied to a range of b_1 extending from negative to positive infinity; within a range of b_1 extending from 0.5 to positive infinity the values of β_1 are expressed with real numbers, while within the remaining range of b_1 with imaginary numbers.

- Kepler's third law of planetary motion is described by Eq. (1.7-2). It is applied to the range of b_1 extending from zero to positive infinity with all values of β_1 expressed with real numbers.

- When $b_1 > 5$, the difference between the values of β_1 calculated based on the toryx law of planetary motion and Kepler's third law of planetary motion becomes small and it decrease as b_1 increases. As b_1 decreases from 5 to 2, this difference progressively increases.

- According to the Kepler's third law of planetary motion, as b_1 decreases from 2 to 0, β_1 sharply increases and approaches positive infinity $(+\infty)$. According to the toryx law of planetary motion, as b_1 decreases from 2 to 0. The value of β_1 initially slightly increases and then, after reaching its maximum value of 1 at $b_1 = 1$, it sharply decreases.

2. Features of Abstract Mathematics of a Toryx

Equations describing the toryx spacetime parameters are mostly based on elementary math commonly taught in high schools. However, to satisfy the toryx spacetime postulates, it is necessary to modify several aspects of elementary math, including the definitions of zero, number line and elementary trigonometric functions. Also, unlike the elementary math that deals with stationary spiral elements, the toryx math considers the spiral elements in motion, explaining its name.

2.1 Infinility versus Elementary Zero

Conventionally, we use the elementary zero (0) in two ways. Firstly, we use it for counting of non-divisible entities. In an elementary number line (Fig. 2.1,1) it appears as an integer immediately preceding number one (1).

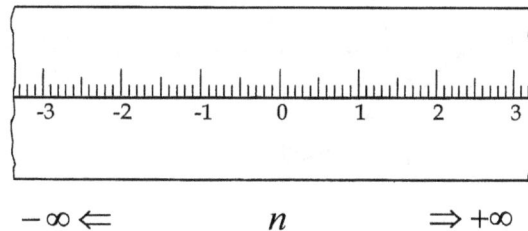

Figure 2.1.1. Elementary number line.

Secondly, we use zero to represent the absolute absence of any quantity and quality. Mathematically, the elementary zero (0) is equal to a ratio of one (1) to infinity (∞). The toryx math clearly separates two applications of zero described above. The zero is still considered as an integer for counting of non-divisible entities and still retains its old symbol (0). But, in application to the spacetime entities the zero is replaced with a quantity that is infinitely approaching to it. This quantity is called **infinility**, from the "infinite nil." (Notably, the term infinility is used in the toryx math instead of the known math term **infinitesimal**). In the toryx math, both infinity and infinility can be positive, negative, real and imaginary as shown below.

$$\text{Real infinility: } \pm 0 = \frac{1}{\pm\infty}; \quad \text{Imaginary infinility: } \pm 0i = \frac{1}{\pm\infty i}$$

$$\text{Real infinity: } \pm \infty = \frac{1}{\pm 0}; \quad \text{Imaginary infinity: } \pm \infty i = \frac{1}{\pm 0i}$$

Figure 2.1.2 shows symbolically positive and negative infinities $(\pm\infty)$ and also positive and negative infinility (± 0) as equal counterparts in respect to the positive and negative unities (± 1).

Infinility $\Leftarrow$ $\Rightarrow$ *Infinity*

(± 0) $(\pm \ddot{1})$ $(\pm\infty)$

Unity

Figure 2.1.2. Infinity $(\pm\infty)$, infinility (± 0) and unity (± 1).

2.2 Toryx Trigonometry

Definitions of elementary trigonometric functions are based on transformations of a right triangle as a function of the non-right angle φ_2 (Fig. 2.2.1):

$$\cos\varphi_2 = x \quad (0^0 < \varphi_2 < 360^0) \tag{2.2-1}$$

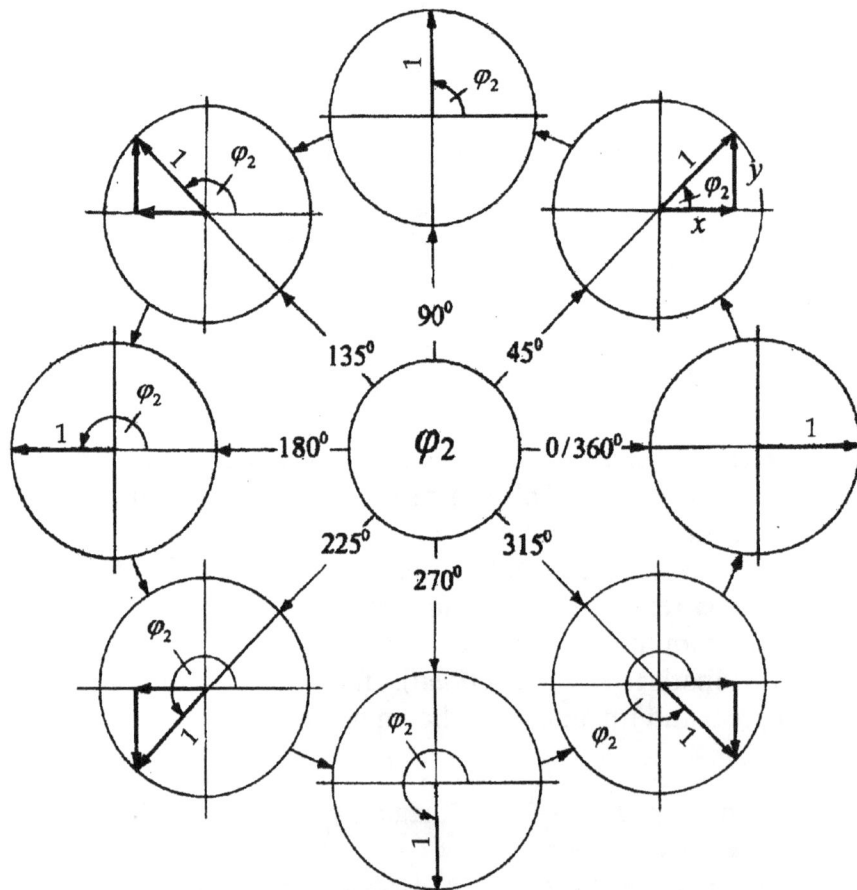

Figure 2.2.1. Transformations of a right triangle in elementary trigonometry.

The main features of the transformations shown in Fig. 2.2.1 are:

- When the length of the hypotenuse of the triangles is equal to 1, the ranges of the lengths of its sides x and y are between 1 and -1.
- The triangles located in two left quadrants are the mirror images of the triangles located in two right quadrants.
- The triangles located in two bottom quadrants are the mirror images of the triangles located at two top quadrants.

In the toryx trigonometry, the transformations of the right triangle are partially modified to satisfy the toryx spacetime postulates. Consequently, the toryx trigonometric function $\cos s\varphi_2$ relates to the elementary trigonometric function $\cos\varphi_2$ as follows:

$$\cos s\varphi_2 = \cos\varphi_2 \quad (0 < \varphi_2 < 180^0) \tag{2.2-2}$$

$$\cos s\varphi_2 = 1/\cos\varphi_2 \quad (180^0 < \varphi_2 < 360^0) \tag{2.2-3}$$

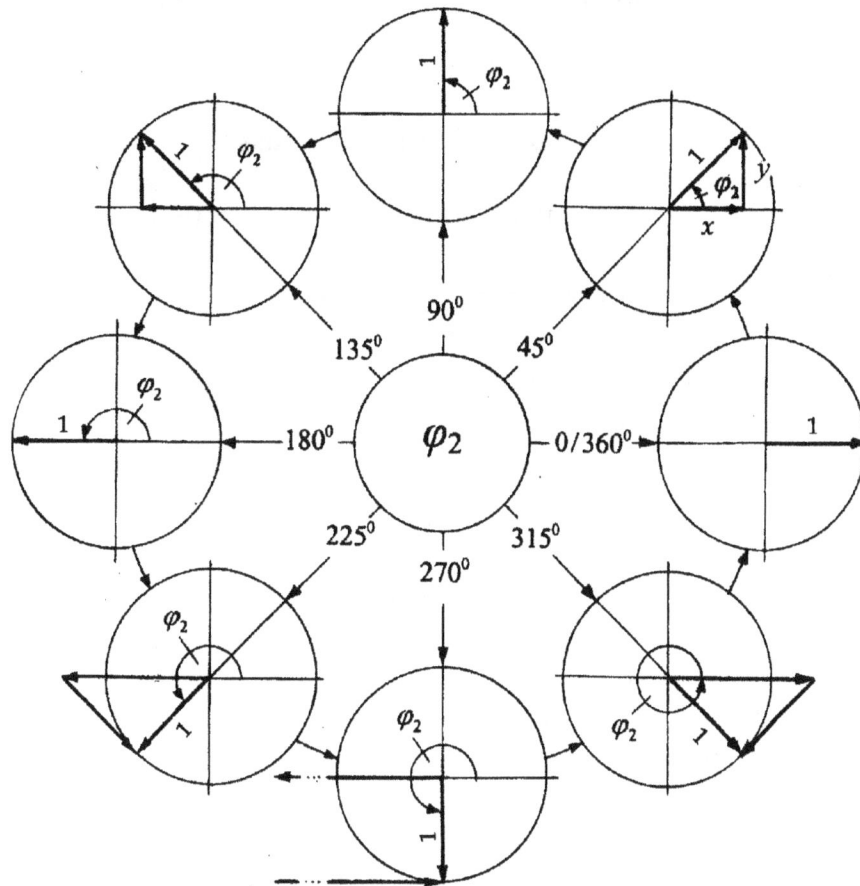

Figure 2.2.2. Transformations of a right triangle in the toryx trigonometry.

The main features of the transformations shown in Fig. 2.2.2 are:

- When the angle φ_2 is between 0 and 180^0, the right triangles are the same as in elementary trigonometry. Thus, within this range of the angle φ_2 the elementary and toryx trigonometry are based on the same principle.
- When the angle φ_2 is between 180 and 360^0, the right triangle becomes ***outverted***. Consequently, the length of its horizontal side x becomes greater than 1, while the length of the other side y is expressed with imaginary numbers.
- When the angle φ_2 approaches 270^0 from the angle smaller than 270^0, the length of its horizontal side x approaches real positive infinity $(+\infty)$, while the length of the other side y approaches imaginary positive infinity $(+\infty i)$.
- When the angle φ_2 approaches 270^0 from the angle greater than 270^0, the length of its horizontal side x approaches real negative infinity $(-\infty)$, while the length of the other side y approaches imaginary negative infinity $(-\infty i)$.
- When the angle φ_2 approaches 360^0 from the angle smaller than 360^0, the length of its horizontal side x approaches 1, while the length of the other imaginary side y approaches imaginary negative infinility $(-0i)$.

2.3 Toryx Number Lines

We consider below four kinds of toryx number lines that are directly related to the toryx parameters:

- *Toryx vorticity V* number line
- *Toryx reality R* number line
- *Toryx boundary B* number line
- *Toryx golden G* number line
- *Toryx string period ratio P* number line.

All five number lines are presented in the forms of circular diagrams in which the numbers V, R, B, S and P are expressed as functions of the steepness angle of toryx trailing string φ_2.

Toryx vorticity V number line - In the toryx vorticity V number line (Fig. 2.3.1), the real numbers V are equal to the ratio of toryx trailing string radius r_2 to the toryx leading string radius r_1 with an opposite sign. These numbers are extended clockwise along a circle from the real positive infinity $(+\infty)$ to the real negative infinity $(-\infty)$ as a function of the steepness angle of trailing string φ_2.

$$V = -\frac{r_2}{r_1} = -\cos s\varphi_2 \qquad (2.3\text{-}1)$$

The toryx vorticity V number line is divided into two domains, the *V infinility domain* and the *V infinity domain*, occupying equal sectors of the circular number line.

- The *V infinility domain* occupies two top quadrants; it contains the values of V extending clockwise from the real positive unity $(+1)$ and passing through infinility (± 0) to the real negative unity (-1).

- The *V infinity domain* resides in two bottom quadrants; it contains the values of V extending counterclockwise from the real positive unity $(+1)$ and passing through infinity $(\pm\infty)$ to the real negative unity (-1).

In the toryx vorticity V number line, the real positive infinility $(+0)$ merges with real negative infinility (-0) at $\varphi_2 = 90^0$, while real negative infinity $(-\infty)$ merges with real positive infinity $(+\infty)$ at $\varphi_2 = 270^0$.

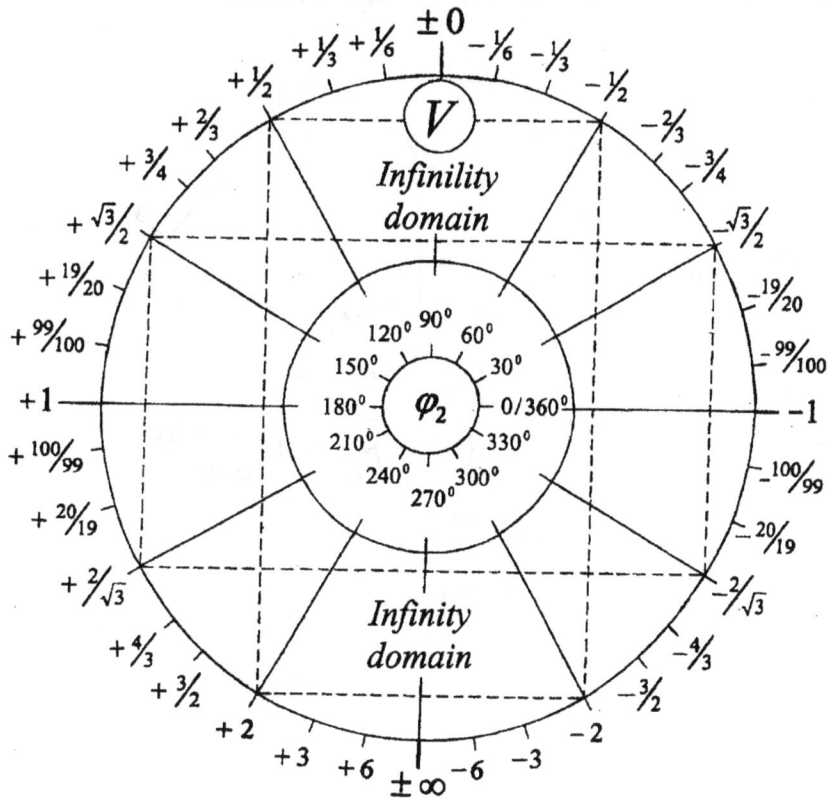

Figure 2.3.1. The toryx vorticity V number line.

There are two kinds of symmetries between the numbers V that belong to the four quadrants of circular diagram, the *inverse V-symmetry* and the *reverse V-symmetry*.

- In the inverse V-symmetry, the magnitudes of the numbers V located in the top quadrants are inversed (reciprocated) in respect to the magnitudes of the numbers V located in the bottom quadrants.

- In the reverse V-symmetry, the numbers V located in the right quadrants and the left quadrants have the same magnitudes but reversed signs.

Toryx reality R number line - In the toryx reality R number line (Fig. 2.3.2), the real and imaginary numbers R are equal to the ratio of toryx trailing string wavelength λ_2 to the circular length of toryx eye $l_0 = 2\pi r_0$. These numbers are extended counterclockwise along a circle from the real positive infinity $(+\infty)$ to the imaginary negative infinity $(-\infty i)$ as a function of the steepness angle of trailing string φ_2.

$$R = \frac{\lambda_2}{2\pi r_0} = \sqrt{\frac{1 + \cos s\varphi_2}{1 - \cos s\varphi_2}} \tag{2.3-2}$$

The toryx reality R number line is divided into two domains, the **R infinility domain** and the **R infinity domain,** occupying equal sectors of the circular number line.

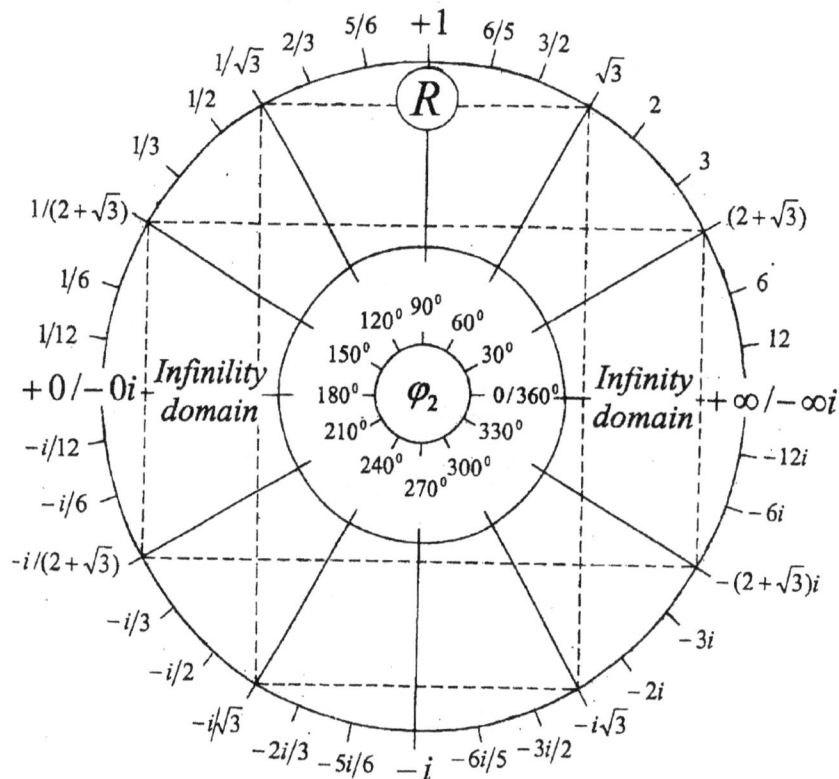

Figure 2.3.2. The toryx reality R number line.

- The R infinility domain occupies two left quadrants; it contains the values of R extending counterclockwise from the real positive unity $(+1)$ and passing through infinility $(+0/0i)$ to the imaginary negative unity $(-i)$.

- The R infinility domain resides in two right quadrants; it contains the values of R extending clockwise from the real positive unity $(+1)$ and passing through real positive and imaginary negative infinities $(+\infty/-\infty i)$ to the imaginary negative unity $(-i)$.

In the toryx reality R number line, the real positive infinility $(+0)$ merges with the imaginary negative infinility $(-0i)$ at $\varphi_2 = 180^0$, while real positive infinity $(+\infty)$ merges with imaginary negative infinity $(-\infty i)$ at $\varphi_2 = 360^0$.

There are two kinds of symmetries between the numbers R that belong to the four quadrants of circular diagram, the ***inverse R-symmetry*** and the ***reverse reality R-symmetry***.

- In the inverse R-symmetry, the magnitudes of the numbers R located in the left quadrants are inversed (reciprocated) in respect to the magnitudes of the numbers R located in the right quadrants.
- In the reverse reality R-symmetry, the numbers R located in the top quadrants are real positive, while these numbers in the bottom quadrants are imaginary negative.

Toryx boundary B number line - In the toryx boundary B number line (Fig. 2.3.3), the real numbers B are equal to the ratio of the radius of toryx radius of spherical boundary r to the toryx eye radius r_0.

$$B = \frac{r}{r_0} = \frac{1 + \cos s\varphi_2}{1 - \cos s\varphi_2} \qquad (2.3\text{-}3)$$

These numbers are extended counterclockwise along a circle from the real positive infinity $(+\infty)$ to the real negative infinity $(-\infty)$ as a function of the steepness angle of trailing string φ_2.

The toryx boundary B number line is divided into two domains, the ***B infinility domain*** and the ***B infinity domain***, occupying equal sectors of the circular number line.

- The B infinility domain occupies two left quadrants; it contains the values of B extending counterclockwise from the real positive unity $(+1)$ and passing through infinility $(+0/-0)$ to the real negative unity (-1).
- The B infinity domain resides in two right quadrants; it contains the values of B extending clockwise from the real positive unity $(+1)$ and passing through real positive and negative infinities $(+\infty/-\infty)$ to the real negative unity (-1).

In the toryx boundary B number line, real positive infinility $(+0)$ merges with real negative infinility (-0) at $\varphi_2 = 180^0$, while real positive infinity $(+\infty)$ merges with real negative infinity $(-\infty)$ at $\varphi_2 = 360^0$.

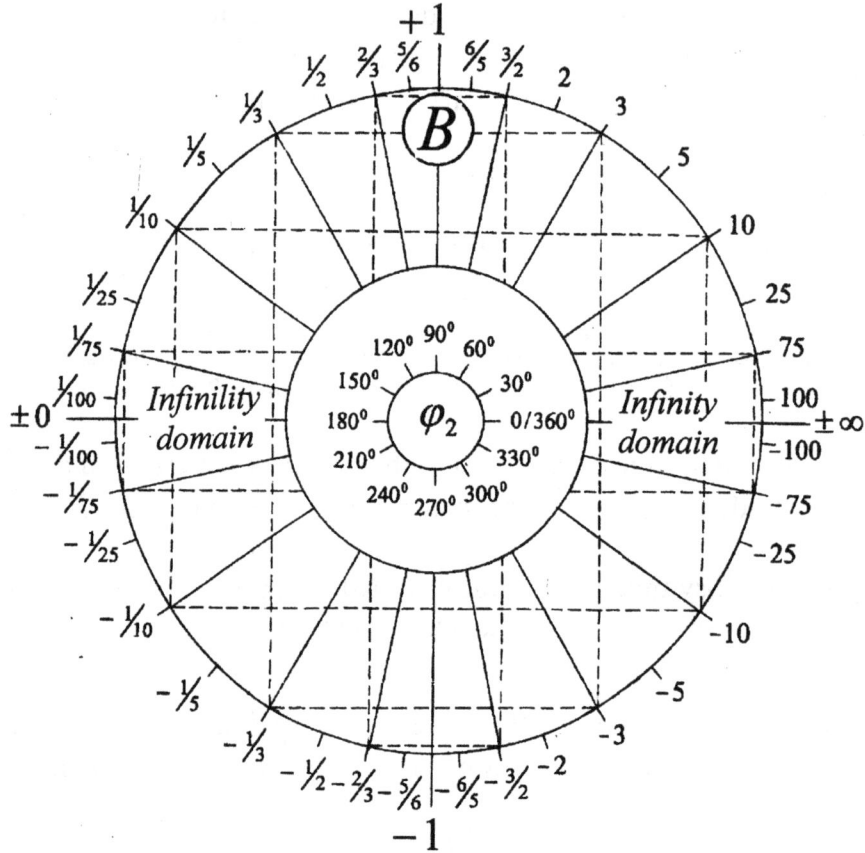

Figure 2.3.3. The toryx boundary B number line.

There are two kinds of symmetries between the numbers B that belong to the four quadrants of circular diagram, the *inverse B-symmetry* and the *reverse B-symmetry*.

- In the inverse B-symmetry, the magnitudes of the numbers B located in the left quadrants are inversed (reciprocated) in respect to the magnitudes of the numbers B located in the right quadrants.
- In the reverse B-symmetry, the numbers B located in the top and bottom quadrants have the same magnitudes but reversed signs.

Toryx golden G number line – In the toryx golden G number line (Fig. 2.3.4), the numbers G are related to the toryx radius of spherical boundary r, the eye radius r_0, the radius of leading string r_1 and the radius of trailing string r_2 of a toryx by the equation:

$$G = -\frac{r\,r_0}{r_1 r_2} = -\frac{1-\cos s^2\varphi_2}{\cos s\varphi_2} \qquad (2.3\text{-}4)$$

When $G = \pm 1$, $\cos s\varphi_2$ relates to the *golden ratio* $\phi = 1.618033989$ as shown in Table 2.3.

Table 2.3. $\cos s\varphi_2$ versus golden ratio ϕ.

Steepness angle of trailing string φ_2	$\cos s\varphi_2$	G
51.83^0	$+1/\phi$	-1
128.17^0	$-1/\phi$	$+1$
231.83^0	$-1/\phi$	-1
308.17^0	$+1/\phi$	$+1$

The toryx golden G number line is divided into four domains, two **G-infinity domains** and two **G-infinility domains**.

- The G-infinity domains occupy top and bottom equal segments of the circular diagram. In both top and bottom G-infinity domains, the values of G extend clockwise from the real positive unity $(+1)$ and passing through infinity $(\pm\infty)$ to the real negative unity (-1).
- The G-infinility domains occupy left and right equal segments of the circular diagram. In both left and right G-infinility domains, the values of G extend clockwise from the real negative unity (-1) and passing through infinility (± 0) to the real positive unity $(+1)$.

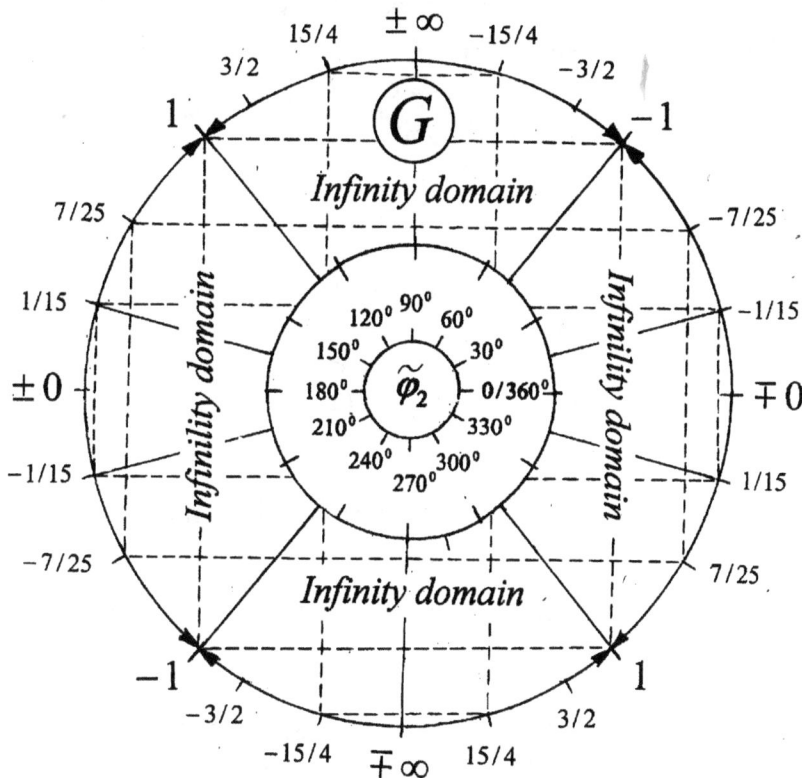

Figure 2.3.4. The toryx golden G number line.

The numbers located in each of four quadrants of the G number line relate to the numbers of adjacent quadrants according to the **reverse G-symmetry**. It means that the numbers G located <u>in each of four quadrants</u> have the same magnitudes but reversed signs in respect to the numbers G located in their adjacent quadrants.

Toryx string period ratio P number line – In the toryx string period ratio P number line (Fig. 2.3.5), the numbers P are related to the ratio of the periods of toryx trailing string T_2 and leading string T_1 by the equation:

$$P = \frac{T_2}{T_1} = \sin s\varphi_2$$

(2.3-5)

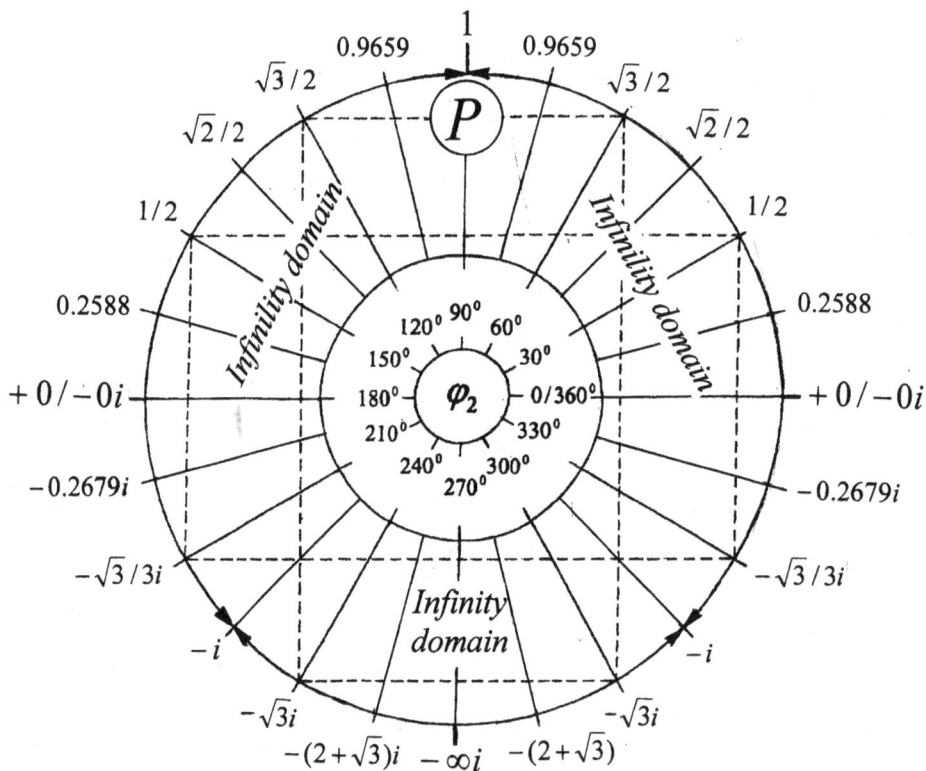

Figure 2.3.5. The toryx string period ratio P number line.

The toryx string period ratio P number line is divided into three domains, two **P-infinility domains** and one **P-infinity domains**.

- <u>Two P-infinility domains</u> occupy top left and right equal segments of the circular diagram. In both left and right top segments, the values of P extend respectively counter-clockwise and clockwise from the real positive unity $(+1)$ and passing through infinility $(+0/-0i)$ to the imaginary negative unity $(-i)$. The numbers P located in the left and right top segments are equal to one another.

- <u>The *P*-infinity domain</u> occupy bottom left and right segments of the circular diagram. In both left and right bottom segments, the values of P extend from the imaginary negative unity $(-i)$ and pass through imaginary negative infinity $(-\infty i)$ to the imaginary negative unity $(-i)$. The numbers P located in the left and right bottom segments are equal to one another.

2.4. Toryx Parameters in Number Lines & Trigonometry

The numbers V of the toryx vorticity V number line (Fig. 2.3.1) relate to the toryx parameters by the equation:

$$V = -\frac{r_2}{r_1} = -\frac{b_2}{b_1} = -\frac{b_1 - 1}{b_1} = -\frac{b - 1}{b + 1} = -\beta_{2r} = -\cos s\varphi_2 \qquad (2.4\text{-}1)$$

The numbers R of the toryx reality R number line (Fig. 2.3.2) relate to the toryx parameters by the equation:

$$R = \eta_2 = \sqrt{b} = \sqrt{2b_1 - 1} = b_1\beta_{2t} = \sqrt{\frac{1 + \cos s\varphi_2}{1 - \cos s\varphi_2}} \qquad (2.4\text{-}2)$$

The numbers B of the toryx boundary B number line (Fig. 2.3.3) relate to the toryx parameters by the equation:

$$B = b = \frac{r}{r_0} = 2b_1 - 1 = \frac{1 + \cos s\varphi_2}{1 - \cos s\varphi_2} \qquad (2.4\text{-}3)$$

The numbers G of the toryx golden G number line (Fig. 2.3.4) are related to the toryx parameters by the equation:

$$G = -\frac{r\,r_0}{r_1\,r_2} = -\frac{b}{b_1\,b_2} = -\frac{4b}{b^2 - 1} = -\frac{2b_1 - 1}{b_1(b_1 - 1)} = -\frac{\beta_{2t}^2}{\beta_{2r}} = -\frac{1 - \cos s^2\varphi_2}{\cos s\varphi_2} \qquad (2.4\text{-}4)$$

When $G = \pm 1$, the toryx parameters relate to the golden ratio ϕ as shown in Table 2.4.

Table 2.4. Relationships between toryx parameters and the golden ratio ϕ when $G = \pm 1$.

φ_2	G	V	b_2	b_1	b
51.83^0	-1	$-1/\phi$	ϕ	$1 + \phi$	$1 + 2\phi$
128.17^0	$+1$	$+1/\phi$	$-1/\phi^2$	$1/\phi$	$2/\phi - 1$
231.83^0	-1	$+1/\phi$	$-1/\phi$	$1/\phi^2$	$2/\phi^2 - 1$
308.17^0	$+1$	$-1/\phi$	$-(1 + \phi)$	$-\phi$	$-(1 + 2\phi)$

The numbers P of the toryx string period ratio P number line relate to the toryx parameters by the equation:

$$P = \frac{T_2}{T_1} == \frac{t_2}{t_1} = \frac{\sqrt{2b_1 - 1}}{b_1} = \frac{1}{w_2} = \beta_1 = \beta_{1r} = \beta_{2t} = \sin s\varphi_2 \qquad (2.4\text{-}5)$$

Figure 2.4 shows the application of the spiral spacetime math for the calculation of the relative velocities of the toryx trailing string corresponding to the middle point of the trailing string as its steepness angle φ_2 increases from 0 to 360^0. In each right triangle of velocities of trailing string one side represents the relative translational velocity β_{2t} and the other side the relative rotational velocity β_{2r}, while its hypotenuse represents the relative spiral velocity $\beta_2 = 1$.

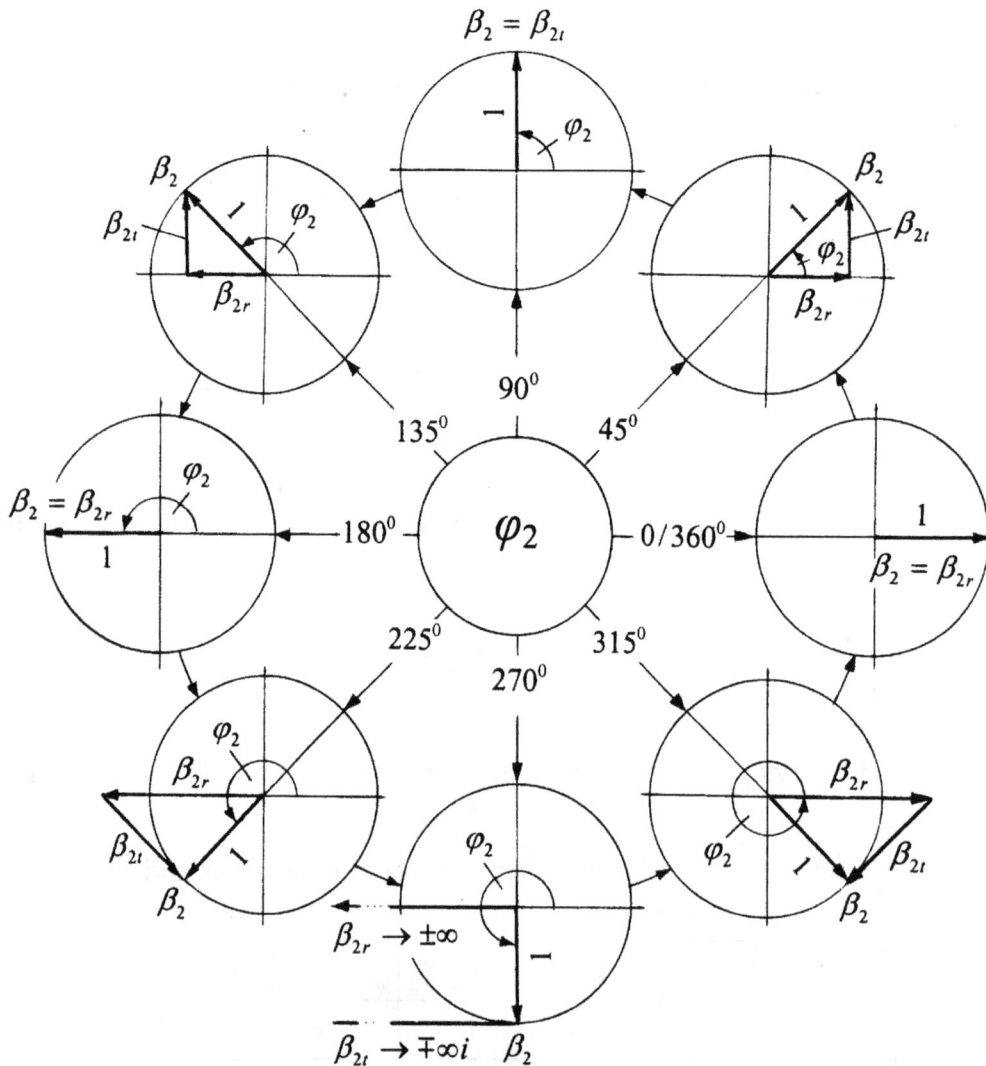

Figure 2.4. Transformations of right triangle representing vectors of the relative velocities of toryx trailing string β_2, β_{2t} and β_{2r} corresponding to the middle point a of trailing string (Fig. 1.1.3).

Notes:

Notes:

3. CLASSIFICATION OF TORYCES

CONTENTS

3.1 Main Groups and Subgroups of Toryces

Depending on the toryx vorticity V and reality R numbers the toryces are divided into four *main groups* and eight **subgroups** as shown in Figure 3.1 and Tables 3.1.1 and 3.1.2.

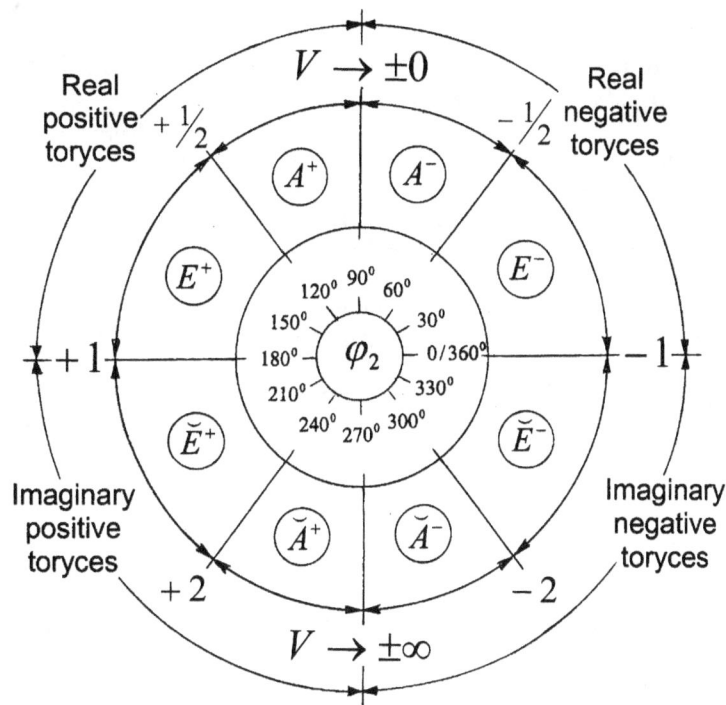

Figure 3.1. Main groups and subgroups of toryces.

The toryces of the main groups are described below and their features are summarized in Table 3.1.1.

Real negative toryces $(0^0 < \varphi_2 < 90^0)$ – The real negative toryces are located in the top right quadrants of the circular diagrams. In the middle point a of trailing strings of these toryces (Fig. 1.1.3), both the translational velocity β_{2t} and the rotational velocity β_{2r} are **subluminal** and expressed with *real numbers*.

Real positive toryces $(90^0 < \varphi_2 < 180^0)$ – The real positive toryces are located in the top left quadrants of the circular diagrams. In the middle point a of trailing strings of these toryces, both the translational velocity β_{2t} and the rotational velocity β_{2r} are **subluminal** and expressed with **real numbers**.

Imaginary positive toryces $(180^0 < \varphi_2 < 270^0)$ – The imaginary positive toryces are located in the bottom left quadrants of the circular diagrams. In the middle point a of trailing strings of these toryces, the rotational velocities β_{2r} are **superluminal** and expressed with **real numbers**, whereas the translational velocities β_{2t} are expressed with **imaginary numbers**.

Imaginary negative toryces $(270^0 < \varphi_2 < 360^0)$ – The imaginary negative toryces are located in the bottom right quadrants of the circular diagrams. In middle point a of trailing strings of these toryces, the rotational velocities β_{2r} are **superluminal** and expressed with **real numbers**, whereas the translational velocities β_{2t} are expressed with **imaginary numbers**.

Table 3.1.1. The realities R, the vorticities V and the mid relative rotational velocities β_{2r} of trailing strings of toryces of main groups.

Toryx name	φ_2	R	V	β_{2r}
Real negative	0^0 - 90^0	Real	(–)	< 1 (subluminal)
Real positive	90^0 - 180^0	Real	(+)	< 1 (subluminal)
Imaginary positive	180^0 - 270^0	Imaginary	(+)	> 1 (superluminal)
Imaginary negative	270^0 - 360^0	Imaginary	(–)	>1 (superluminal)

Within each main group, the toryces are further divided into two **subgroups** as shown in Fig. 3.1 and Table 3.1.2.

Table 3.1.2. Toryces of main groups and subgroups.

Toryces of main groups	Toryces of subgroups	
	Name	φ_2
Real negative	E^-	$0^0 < \varphi_2 < 60^0$
	A^-	$60^0 < \varphi_2 < 90^0$
Real positive	A^+	$90^0 < \varphi_2 < 120^0$
	E^+	$120^0 < \varphi_2 < 180^0$
Imaginary positive	$\breve{E}^+$	$180^0 < \varphi_2 < 240^0$
	$\breve{A}^+$	$240^0 < \varphi_2 < 270^0$
Imaginary negative	$\breve{A}^-$	$270^0 < \varphi_2 < 300^0$
	$\breve{E}^-$	$300^0 < \varphi_2 < 360^0$

3.2 Vorticities & Realities of Adjacent Toryces

Adjacent toryces of main groups – Table 3.2.1 shows relationships between the vorticities V and the realities R of adjacent toryces of main groups.

Table 3.2.1. Relationships between the vorticity V and the reality R of adjacent toryces of main groups.

Adjacent toryces of main groups		Eqs. (a)	Eqs. (b)
Reality-polarized negative toryces	$\breve{E}^- \,\&\, E^-$ Eq. (3.2-1)	$\breve{V}_E^- = 1/V_E^-$	$\breve{R}_E^- = \pm i R_E^-$
Reality-polarized positive toryces	$\breve{E}^+ \,\&\, E^+$ Eq. (3.2-2)	$\breve{V}_E^+ = 1/V_E^+$	$\breve{R}_E^+ = \pm i R_E^+$
Vorticity-polarized real toryces	$A^+ \,\&\, A^-$ Eq. (3.2-3)	$V_A^+ = -V_A^-$	$R_A^+ = 1/R_A^-$
Vorticity-polarized imaginary toryces	$\breve{A}^+ \,\&\, \breve{A}^-$ Eq. (3.2-4)	$\breve{V}_A^+ = -\breve{V}_A^-$	$\breve{R}_A^+ = 1/\breve{R}_A^-$

Table 3.2.2 shows relationships between the relative radii of leading strings b_1 and spherical boundary b of adjacent toryces of main groups.

Table 3.2.2. Relationships between the relative radii of leading string b_1 and spherical boundary b of adjacent toryces of main groups.

Toryces of main groups		Eqs. (a)	Eqs. (b)
Reality-polarized negative toryces	$\breve{E}^- \,\&\, E^-$ Eq. (3.2-5)	$\breve{b}_{1E}^- = 1 - b_{1E}^-$	$\breve{b}_E^- = -b_E^-$
Reality-polarized positive toryces	$\breve{E}^+ \,\&\, E^+$ Eq. (3.2-6)	$\breve{b}_{1E}^+ = 1 - b_{1E}^+$	$\breve{b}_E^+ = -b_E^+$
Vorticity-polarized real toryces	$A^+ \,\&\, A^-$ Eq. (3.2-7)	$b_{1A}^+ = \dfrac{b_{1A}^-}{2b_{1A}^- - 1}$	$b_A^+ = \dfrac{1}{b_A^-}$
Vorticity-polarized imaginary toryces	$\breve{A}^+ \,\&\, \breve{A}^-$ Eq. (3.2-8)	$\breve{b}_{1A}^+ = \dfrac{\breve{b}_{1A}^-}{2b_{1A}^- - 1}$	$\breve{b}_A^+ = \dfrac{1}{\breve{b}_A^-}$

Adjacent toryces of subgroups - Exhibit 3.2 shows a limitation of degrees of freedom for adjacent real negative toryces $A^- \,\&\, E^-$ at the borderline between them. Table 3.2.3 shows the relationships between the vorticities V of adjacent toryces of subgroups.

Exhibit 3.2. The vorticities V of adjacent real negative toryces $A^- \& E^-$
at the borderline between them.

The vorticities V of adjacent real negative toryces $A^- \& E^-$ at the borderline between them are the same and they are equal to:

$$V_A^- = V_E^- = -0.5 \qquad (3.2\text{-}9)$$

Table 3.2.3. Relationships between vorticities V of adjacent toryces of subgroups.

Adjacent toryces of subgroups		Equations
Real negative toryces	$A^- \& E^-$	$V_A^- + V_E^- = -1$ (3.2-10)
Real positive toryces	$A^+ \& E^+$	$V_A^+ + V_E^+ = +1$ (3.2-11)
Imaginary positive toryces	$\breve{A}^+ \& \breve{E}^+$	$\dfrac{1}{V_A^+} + \dfrac{1}{V_E^+} = +1$ (3.2-12)
Imaginary negative toryces	$\breve{A}^- \& \breve{E}^-$	$\dfrac{1}{V_A^-} + \dfrac{1}{V_E^-} = -1$ (3.2-13)

Table 3.2.4 summarizes relationships between the relative radii of leading string b_1 and spherical boundary b of toryces of subgroups.

Table 3.2.4. Relationships between the relative radii of leading string b_1 and spherical boundary b of toryces of subgroups.

Toryces of subgroups		Eqs. (a)	Eqs. (b)
Negative toryces	$A^- \& E^-$ Eq. (3.2-14)	$b_{1A}^- = \dfrac{b_{1E}^-}{b_{1E}^- - 1}$	$b_A^- = \dfrac{b_E^- + 3}{b_E^- - 1}$
Positive toryces	$A^+ \& E^+$ Eq. (3.2-15)	$b_{1A}^+ = \dfrac{b_{1E}^+}{3b_{1E}^+ - 1}$	$b_A^+ = \dfrac{1 - b_E^+}{1 + 3b_E^+}$

3.3 Self-Polarized Toryces

In Section 3.2, we defined the reality-polarized toryces based on their reality R. This reality corresponds to the reality of the translational velocities β_{2t} at the middle points a of toryx trailing strings (Fig. 1.1.3). When β_{2t} is expressed with real numbers the toryx is considered real. Conversely, when β_{2t} is expressed with imaginary numbers the toryx is called imaginary.

Creation of the self-polarized toryces becomes possible, because during each cycle of the instant translational and rotational velocities of their trailing strings vary. As shown in Fig. 3.3.1, the instant translational velocity at each point is proportional to the distance of that point from the toryx center. At the same time, the rotational velocity of trailing string must vary to satisfy Eq. (1.5-3) according to which the trailing string must propagate at a constant relative spiral string velocity $\beta_2 = 1$.

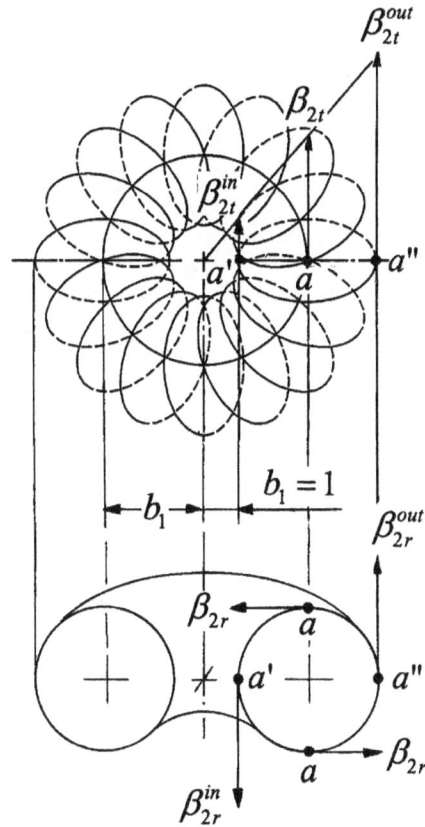

Figure 3.3.1. Relative peripheral velocities of trailing string.

Table 3.3 shows equations for relative inner and outer peripheral translational and rotational velocities of trailing strings.

Table 3.3. Relative inner and outer peripheral velocities of trailing string.

Inner velocities		Outer velocities	
At the point a' (Fig. 3.3.1)		at the point a'' (Fig. 3.3.1)	
$\beta_{2t}^{in} = \dfrac{\sqrt{2b_1 - 1}}{b_1^2}$	(3.3-1)	$\beta_{2t}^{out} = \dfrac{(2b_1 - 1)^{3/2}}{b_1^2}$	(3.3-2)
$\beta_{2r}^{in} = \dfrac{\sqrt{b_1^4 - 2b_1 + 1}}{b_1^2}$	(3.3-3)	$\beta_{2r}^{out} = \dfrac{\sqrt{b_1^4 - (2b_1 - 1)^3}}{b_1^2}$	(3.3-4)

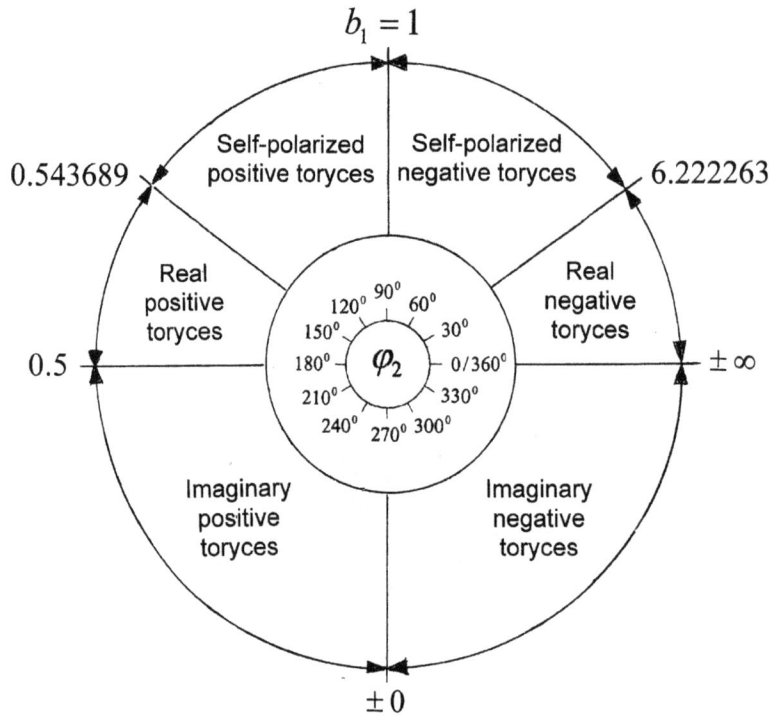

Figure 3.3.2. The ranges of the relative radius of leading string b_1 of self-polarized toryces.

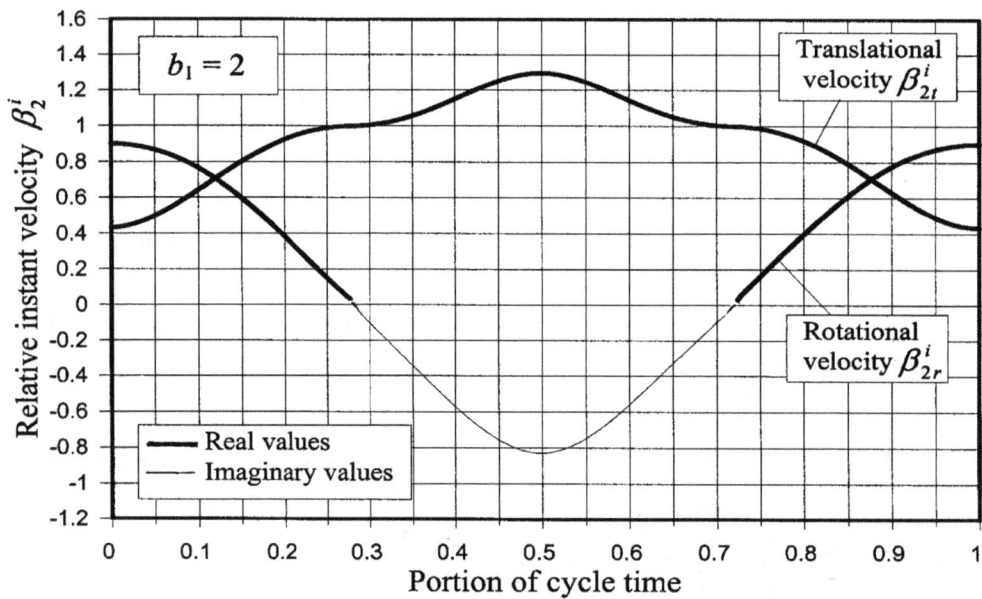

Figure 3.3.3. Variation of instant velocities β_{2t}^i and β_{2r}^i of trailing string within one cycle of trailing string.

The analysis of these equations shows that there are two ranges of the relative radii of toryx leading string b_1 within which the instant rotational velocities of trailing strings change in two different ways during their one cycle (Fig. 3.3.2).

The first range extends from $b_1 = 0.543689$ and then passes through $b_1 \rightarrow \pm 0$ to $b_1 = 6.222263$. Within that range the realities R of the instant rotational velocities of trailing strings remain real during their one cycle.

The second range extends from $b_1 = 0.543689$ and then passes through $b_1 = 1$ to $b_1 = 6.222263$. Within that range the realities R of the instant rotational velocities of trailing strings change from real to imaginary during their one cycle. Consequently, the toryces with b_1 within that range become self-polarized.

Figure 3.3.3 shows the variation of the relative instant translational and rotational velocities, β_{2t}^i and β_{2r}^i for the case when $b_1 = 2$. Notably, between 0.25 and 0.75 portions of the cycle of trailing string, β_{2t}^i exceeds velocity of light, while β_{2r}^i is expressed with imaginary numbers.

<u>*Notes*</u>

4. TRENDS OF TORYX PARAMETERS

CONTENTS

The toryces having the spacetime parameters described in the previous Chapters are called the *basic toryces*. Presented below are the plots of equations expressing the basic toryx parameters as functions of the steepness angle of trailing string φ_2.

4.1 Radii of Leading & Trailing Strings

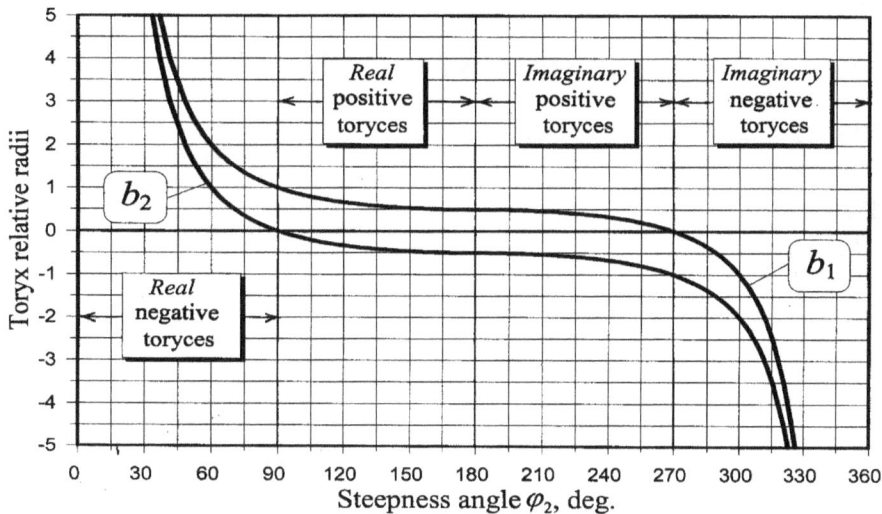

φ_2	$360^0/0^0$	90^0	180^0	270^0
b_1	$-\infty/+\infty$	$+1$	$+\frac{1}{2}$	$+0/-0$
b_2	$-\infty/+\infty$	$+0/-0$	$-\frac{1}{2}$	-1

Figure 4.1. Trends of the relative radii of leading and trailing strings, b_1 and b_2.

4.2 Radius of Spherical Boundary

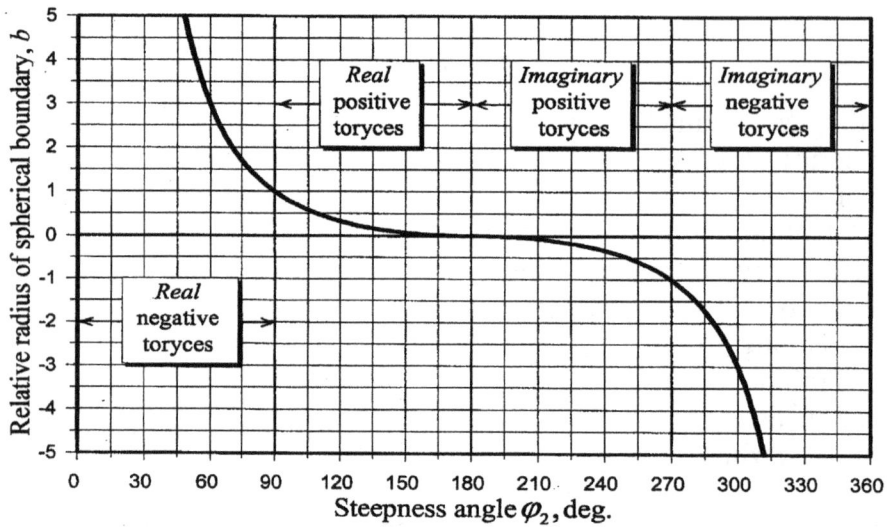

φ_2	$360^0/0^0$	90^0	180^0	270^0
b	$-\infty/+\infty$	$+1$	$+0/-0$	-1

Figure 4.2. Trend of the toryx relative radius of spherical boundary b.

4.3 Wavelength of Trailing String

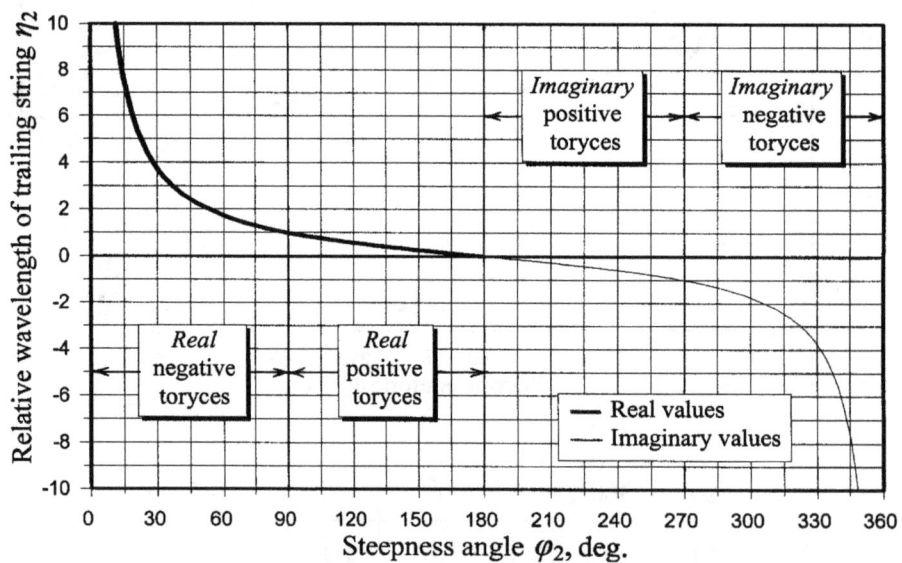

φ_2	$360^0/0^0$	90^0	180^0	270^0
η_2	$-\infty i/+\infty$	$+1$	$+0/-0i$	$-i$

Figure 4.3. Trend of the relative wavelength of trailing string η_2.

4.4 The Number of Windings of Trailing String

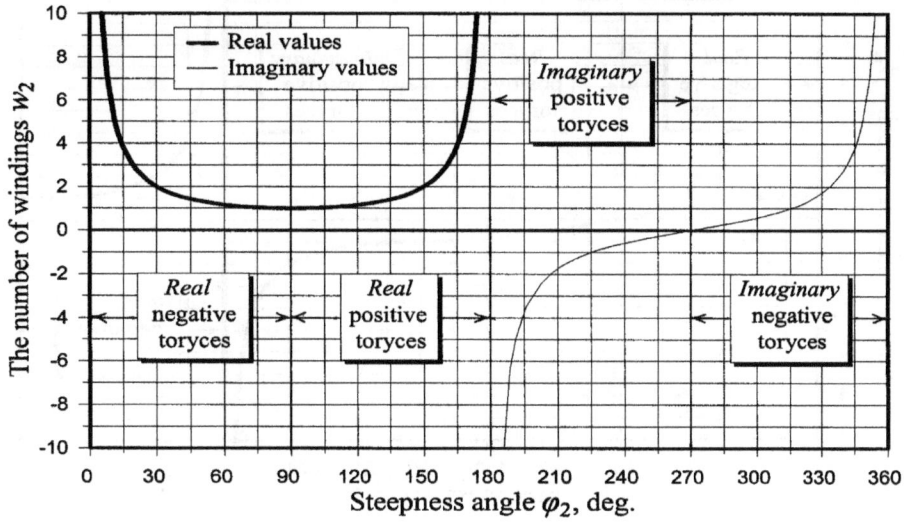

φ_2	$360^0 / 0^0$	90^0	180^0	270^0
w_2	$+\infty i /+\infty$	$+1$	$+\infty /-\infty i$	$-0i /+0i$

Figure 4.4. Trend of the number of windings of trailing string w_2.

4.5 Translational Velocity of Trailing String

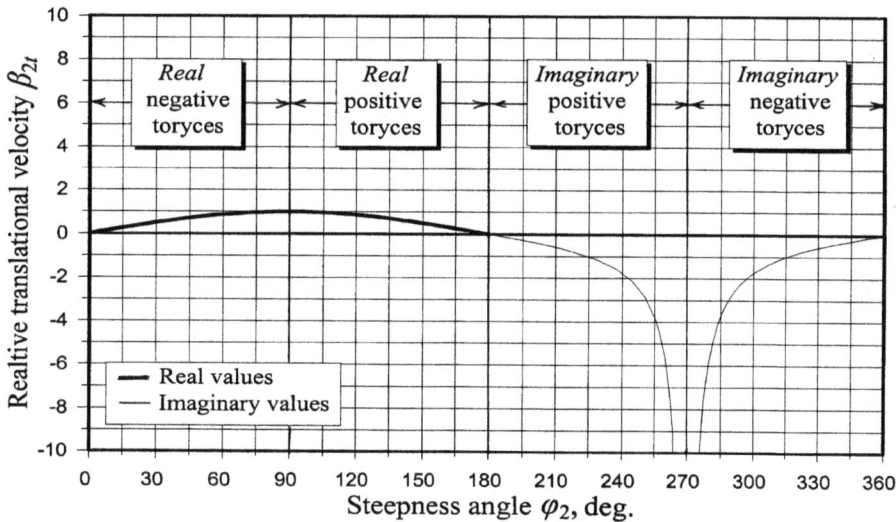

φ_2	$360^0 / 0^0$	90^0	180^0	270^0
β_{2t}	$-0i /+0$	$+1$	$+0 /-0i$	$-\infty i$

Figure 4.5. Trend of the relative translational velocity of trailing string β_{2t}.

4.6 Rotational Velocity of Trailing String

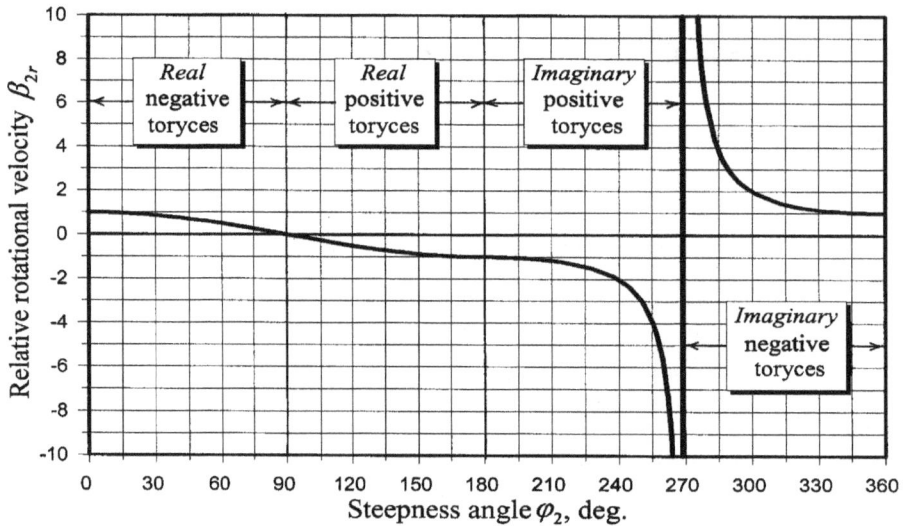

φ_2	$360^0/0^0$	90^0	180^0	270^0
β_{2r}	$+1$	$+0/-0$	-1	$-\infty/+\infty$

Figure 4.6. Trend of the relative rotational velocity of trailing string β_{2r}.

4.7 Frequency of Leading String

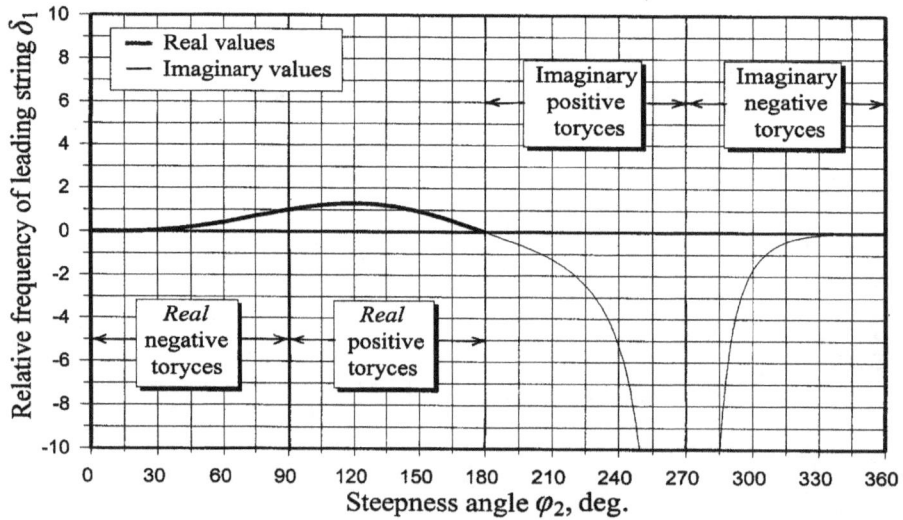

φ_2	$360^0/0^0$	90^0	180^0	270^0
δ_1	$-0i/+0$	$+1$	$+0/-0i$	$-\infty i$

Figure 4.7. Trend of the relative frequency of leading string δ_1.

4.8 Frequency of Trailing String

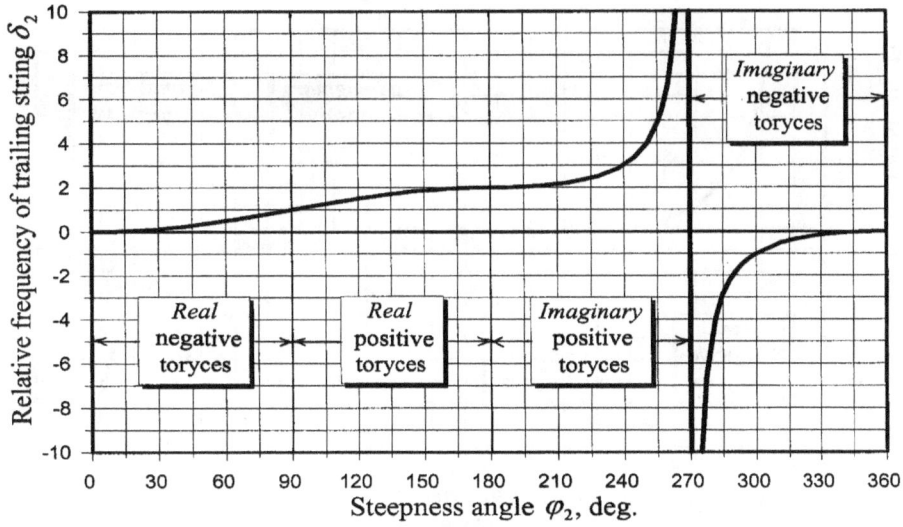

φ_2	$360^0/0^0$	90^0	180^0	270^0
δ_2	$-0/+0$	$+1$	$+2$	$+\infty/-\infty$

Figure 4.8. Trend of the relative frequency of trailing string δ_2.

4.9 Period of Leading String

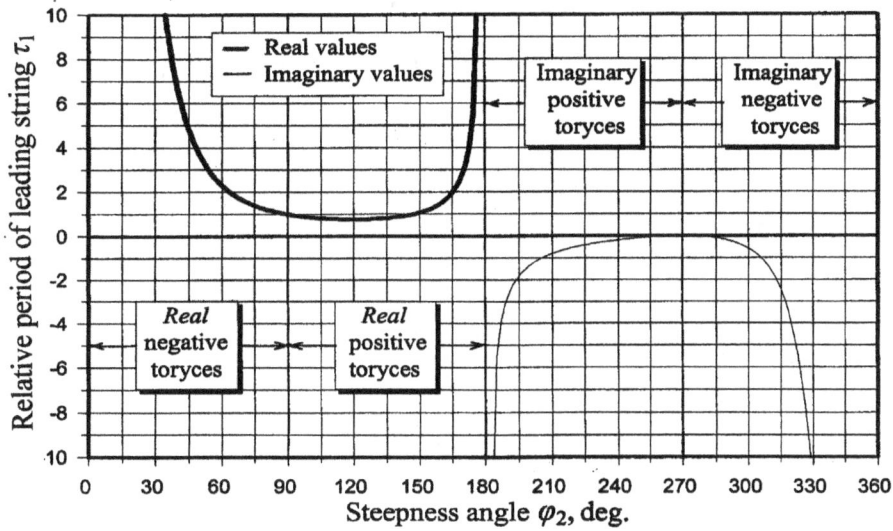

φ_2	$360^0/0^0$	90^0	180^0	270^0
τ_1	$-\infty i/+\infty$	$+1$	$+\infty/-\infty i$	$-0i$

Figure 4.9. Trend of the relative period of leading string τ_1.

4.10 Period of Trailing String

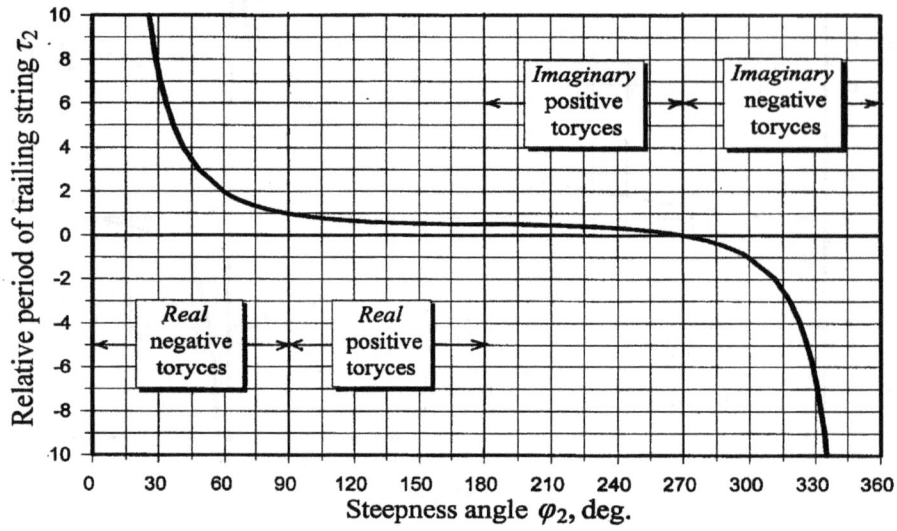

φ_2	$360^0/0^0$	90^0	180^0	270^0
τ_2	$-\infty/+\infty$	$+1$	$+\frac{1}{2}$	$+0/-0$

Figure 4.10. Trend of the relative period of trailing string τ_2.

5. Inversion of Toryces

5.1 Inversion of Toryx Leading & Trailing Strings

As the radius of toryx leading string b_1 decreases from positive to negative infinity, the leading string and trailing string undergo through spacetime transformations that can be visualized as described below.

Inversion of leading string – Visualize the toryx leading string in the form of an extremely thin and narrow circular ribbon with the relative radius b_1 (Fig. 5.1.1). When $b_1 > 0$; the outer color of the circular ribbon is assumed to be black while its inner color is white. The leading string remains **outverted** until b_1 reduces to positive infinility $(+0)$. When b_1 approaches infinility (± 0), the leading string becomes **inverted**. So, when $b_1 < 0$, the outer color of leading string appears white, while its inner color looks black.

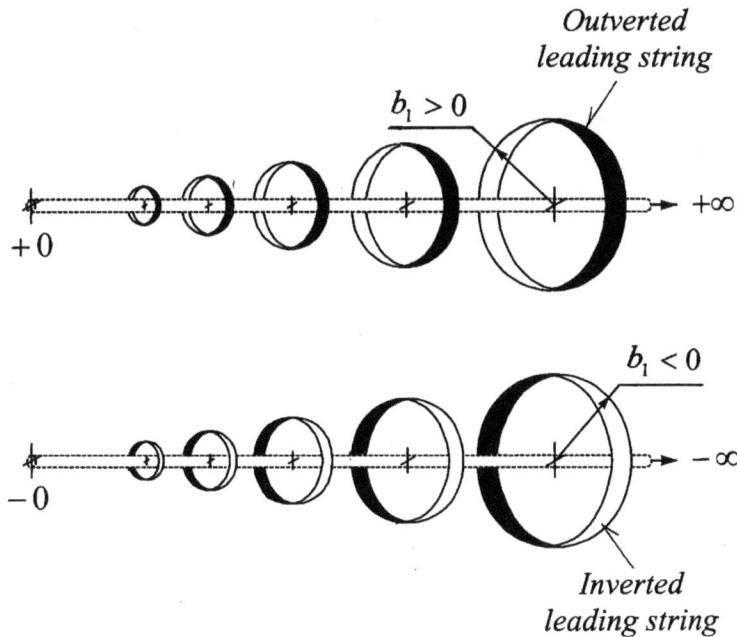

Fig. 5.1.1. Inversion of a leading string.

Inversion of a trailing string - Visualize the toryx trailing string in the form of an extremely thin and narrow toroidal ribbon with the relative radius b_2 (Fig. 5.1.2). When the relative radius of leading string $b_1 > 1$ the radius of trailing string $b_2 > 1$. For that case, the outer color of toroidal ribbon is assumed to be black, while its inner color white. The trailing string remains *outverted* until b_2 reduces to positive infinity $(+0)$. When $b_1 = 1$, b_2 approaches infinility (± 0) and the trailing string becomes *inverted*. So, when $b_2 < 1$, the outer color of leading string appears white, while its inner color looks black.

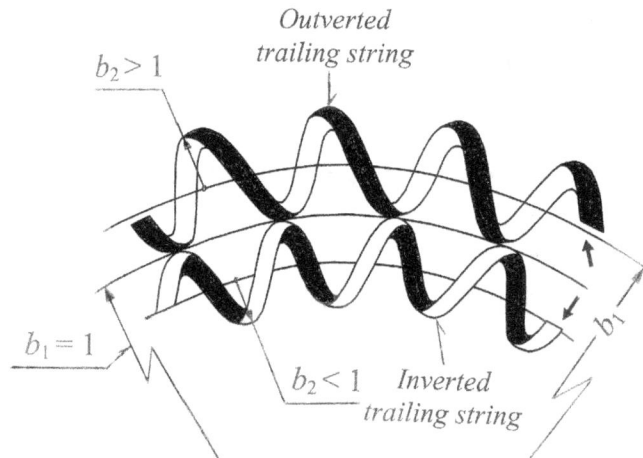

Fig. 5.1.2. Inversion of a trailing string.

Inversion of a spherical boundary - Visualize an extremely thin spherical boundary with the relative radius b (Fig. 5.1.3).

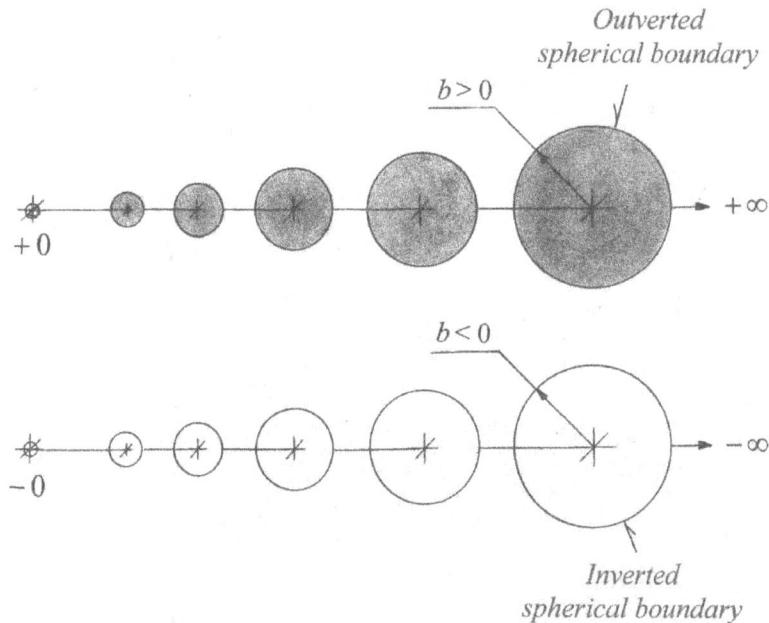

Fig. 5.1.3. Inversion of a spherical boundary.

When $b > 0$, the outer color of the spherical boundary is assumed to be grey and its inner color is assumed to be white. The spherical boundary remains *outverted* until b reduces to positive infinility $(+0)$. When b approaches infinility (± 0), the spherical boundary becomes *inverted*. So, when $b < 0$, the outer color of spherical boundary appears white, while its inner color becomes grey.

5.2 Metamorphoses of Toryces

Figure 5.2 shows that as the radius of the toryx leading string decreases from positive to negative infinity, the steepness angle of trailing string φ_2 increases from 0^0 to 360^0, while the *toryx vorticity* V extends equally from *Unity* (± 1) towards both *Infinility* (± 0) and *Infinity* $(\pm \infty)$.

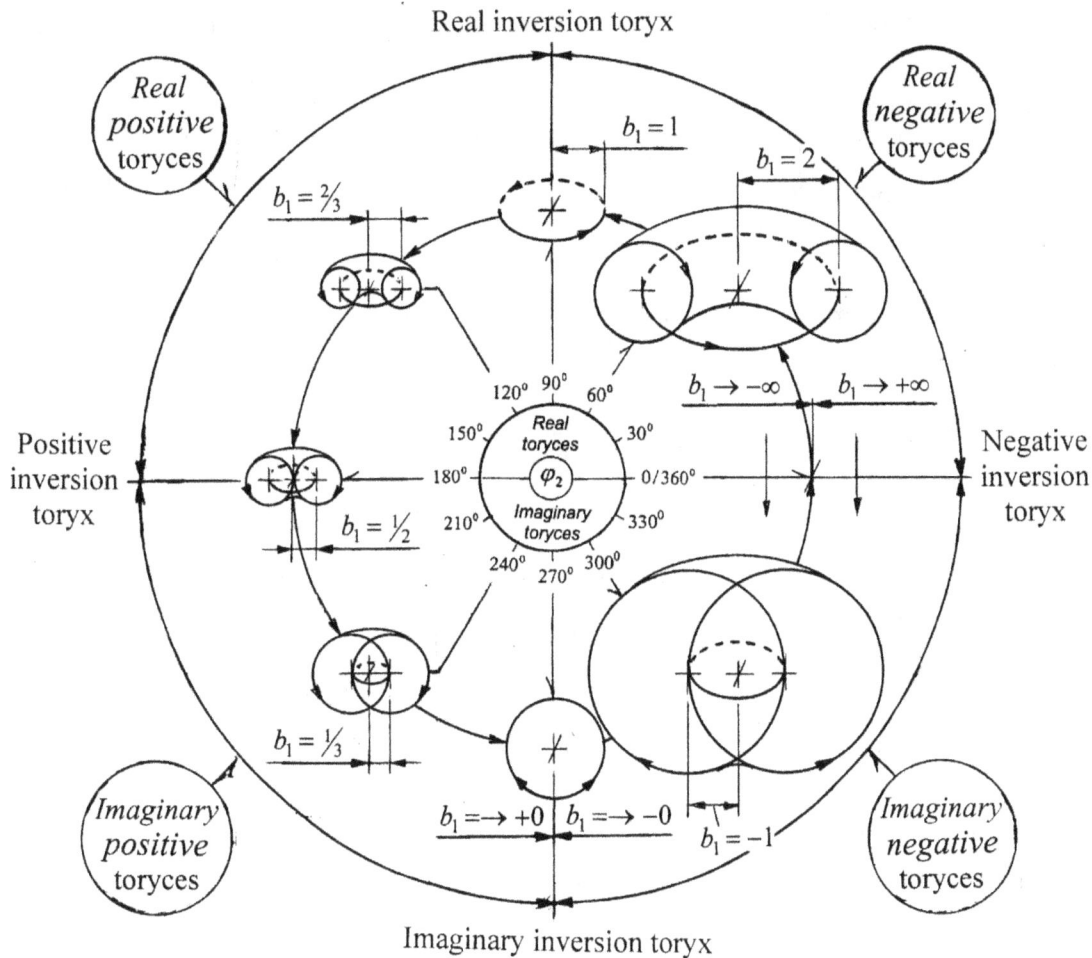

Figure 5.2. Metamorphoses of toryx leading and trailing strings as a function of the steepness angle of trailing string φ_2.

Four kinds of *inversion toryces* are located at the boundaries of four quadrants of the circular diagram.

- **Negative inversion toryx** $(\varphi_2 \to 0^0/360^0)$ - At this point, the toryx leading string, trailing string and its wavelength become inverted, and the toryx appears as two parallel lines separated by the distance equal to the diameter of real inversion string.

- **Real inversion toryx** $(\varphi_2 \to 90^0)$ – At this point, the toryx trailing string becomes inverted $(b_1 \to +1, b_2 \to \pm 0, \eta_2 = 1)$, and the toryx appears as a circle with the relative radius $b_1 \to +1$.

- **Positive inversion toryx** $(\varphi_2 \to 180^0)$ – At this point, the wavelength of toryx trailing string becomes inverted $(b_1 \to +0.5, b_2 \to -0.5, \eta_2 \to +0/-0i)$, and the toryx appears as an extreme case of a *spindle torus* with the inner parts of its windings touching one another.

- **Imaginary inversion toryx** $(\varphi_2 \to 270^0)$ – At this point, the toryx leading string becomes inverted $(b_1 \to \pm 0, b_2 \to -1, \eta_2 = -i)$, and the toryx appears as a circle with the relative radius approaching -1. The circle is located at the plane perpendicular to the plane of the real inversion string.

Table 5.2 shows extreme relative parameters of inversion toryces.

Table 5.2. Extreme relative parameters of inversion toryces.

Inversion toryces	φ_2	b	b_1	b_2	η_2	w_2	β_{2t}	β_{2r}	δ_1	δ_2
Negative inversion toryx	$0^0/360^0$	$-\infty$ $+\infty$	$-\infty$ $+\infty$	$-\infty$ $+\infty$	$-\infty i$ $+\infty$	$+\infty i$ $+\infty$	$-0i$ $+0$	$+0$ -0	$-0i$ $+0$	-0 $+0$
Real inversion toryx	90^0	-	-	$+0$ -0	-	-	-	$+0$ -0	-	-
Positive inversion toryx	180^0	$+0$ -0	$+0$ -0	-	$+0$ $-0i$	$+\infty$ $-\infty i$	$+0$ $-0i$	$-\infty$ $+\infty$	$+0$ $-0i$	-
Imaginary inversion toryx	270^0	-	-	-	-	$-0i$ $+0i$	$-\infty i$	-	$-\infty i$	$+\infty$ $-\infty$

5.3 Transformations of Real Negative Toryces

Real negative toryces (Fig. 5.3) belong to the top right quadrant of the circular diagram shown in Fig. 5.2. Trailing strings of these toryces are wound counter-clockwise outside of the real inversion toryx. As φ_2 increases, both b_1 and b_2 decrease, so that the trailing string appears like a conventional torus.

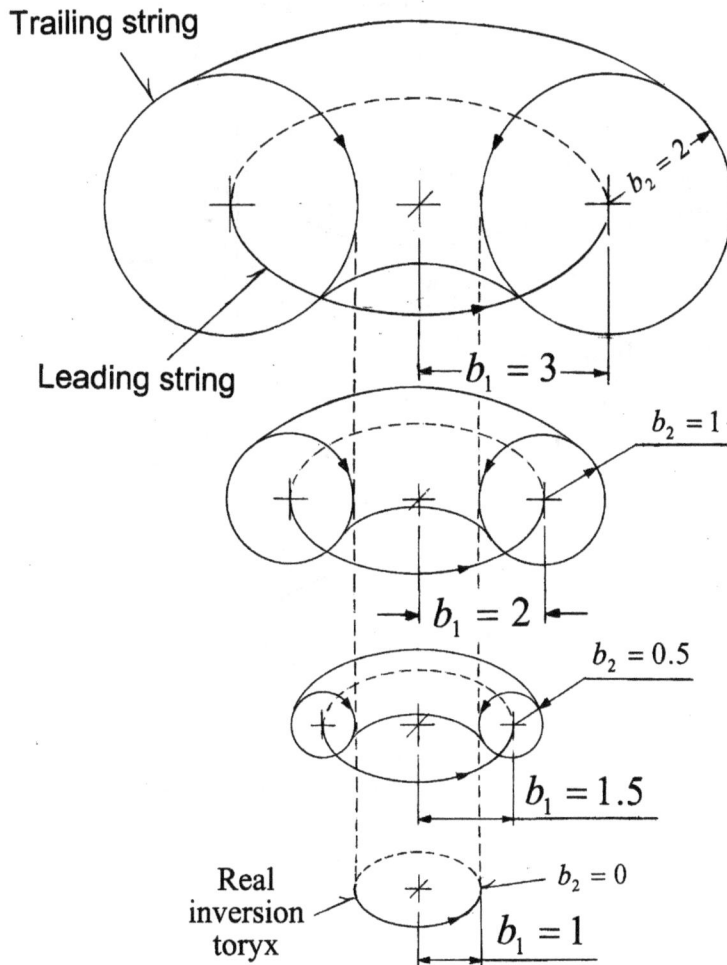

Range	φ_2	b	b_1	b_2	η_2	w_2	β_{2t}	β_{2r}
From	0^0	$+\infty$	$+\infty$	$+\infty$	$+\infty$	$+\infty$	$+0$	1.0
To	90^0	1.0	1.0	$+0$	1.0	1.0	1.0	$+0$

Figure 5.3. Transformations of real negative toryces.

5.4 Transformations of Real Positive Toryces

Real positive toryces (Fig. 5.4) belong to the top left quadrant of the circular diagram shown in Figure 5.2. Within this range the trailing string is inverted, so that its windings are now wound clockwise inside the real inversion toryx. As φ_2 increases, b_1 decreases, while the negative value of b_2 increases. Consequently, the toryx appears as an inverted toroidal spiral.

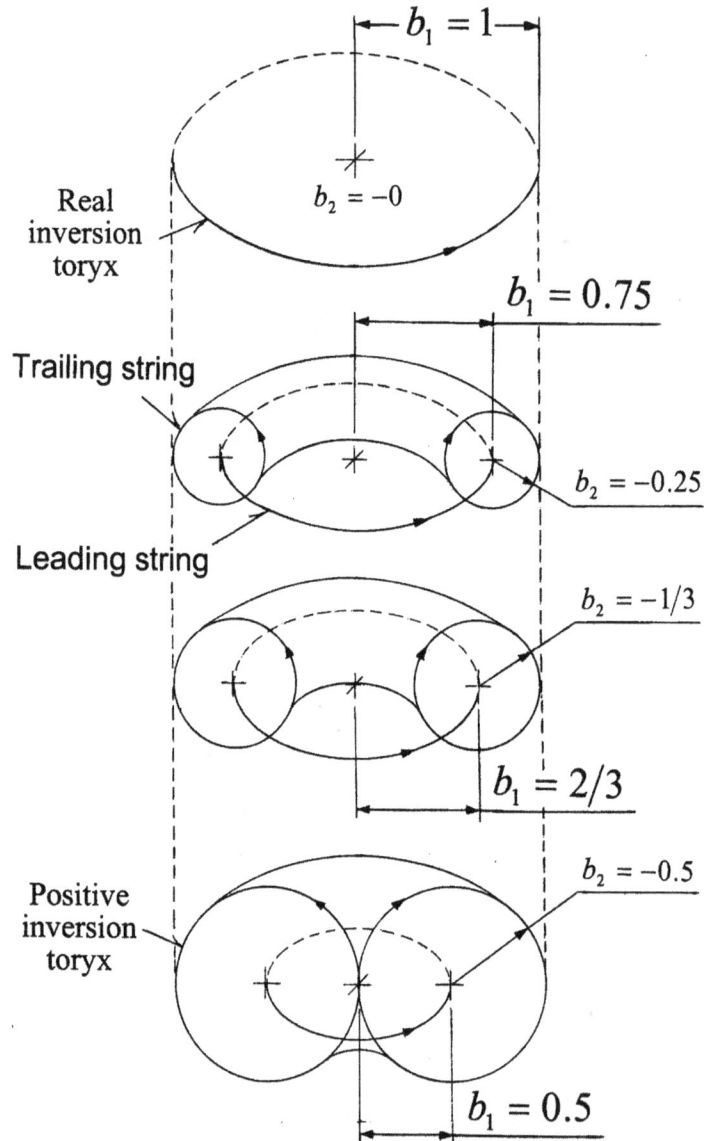

Range	φ_2	b	b_1	b_2	η_2	w_2	β_{2t}	β_{2r}
From	90^0	1.0	1.0	- 0	1.0	1.0	1.0	- 0
To	180^0	+0	0.5	-0.5	+0	$+\infty$	+0	-1.0

Figure 5.4. Transformations of real positive toryces.

5.5 Transformations of Imaginary Positive Toryces

Real positive toryces (Fig. 5.5) belong to the bottom left quadrant of the circular diagram shown in Figure 5.2. As φ_2 increases, b_1 decreases while negative values of b_2 increase. Within this range, the opposite parts of windings of trailing string intersect with one another like in a spindle torus.

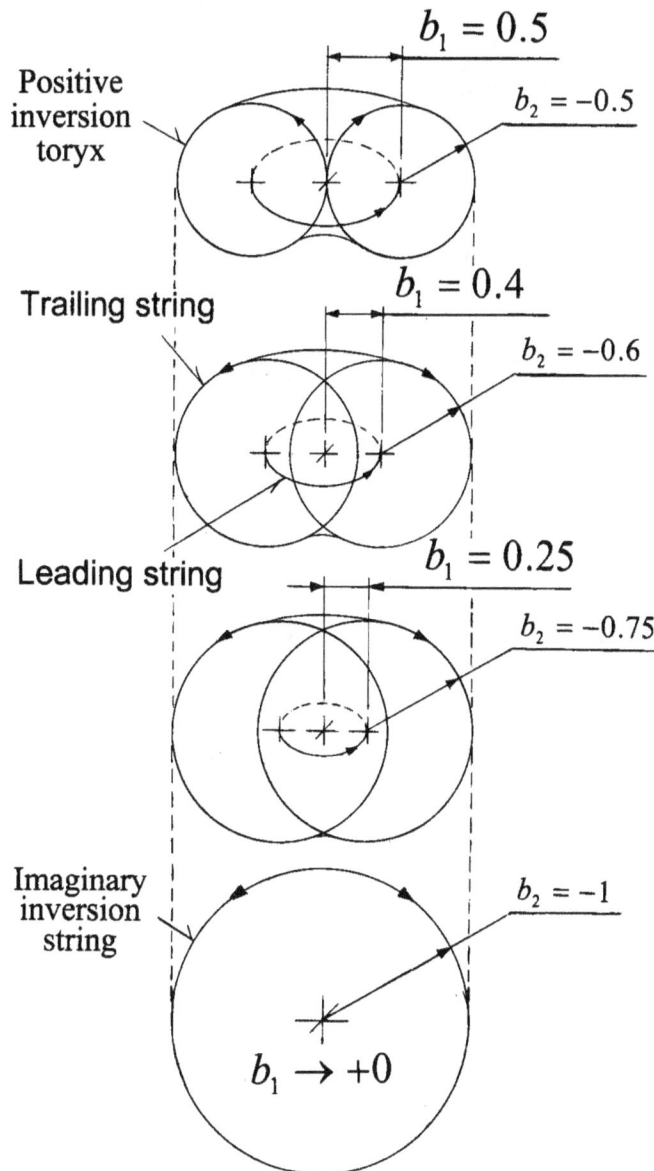

Figure 5.5. Transformations of imaginary positive toryces.

Range	φ_2	b	b_1	b_2	η_2	w_2	β_{2t}	β_{2r}
From	180^0	-0	0.5	-0.5	$-0i$	$-\infty i$	$-0i$	-1.0
To	270^0	-1.0	$+0$	-1.0	$-i$	$-0i$	$-\infty i$	$-\infty$

5.6 Transformations of Imaginary Negative Toryces

Imaginary negative toryces (Fig. 5.6) belong to the bottom right quadrant of the circular diagram shown in Figure 5.2. Here the leading string becomes inverted. As φ_2 increases, the negative values of b_1 increase, the negative values of b_2 also increase. Within this range the toryx windings are located outside of the imaginary inversion toryx.

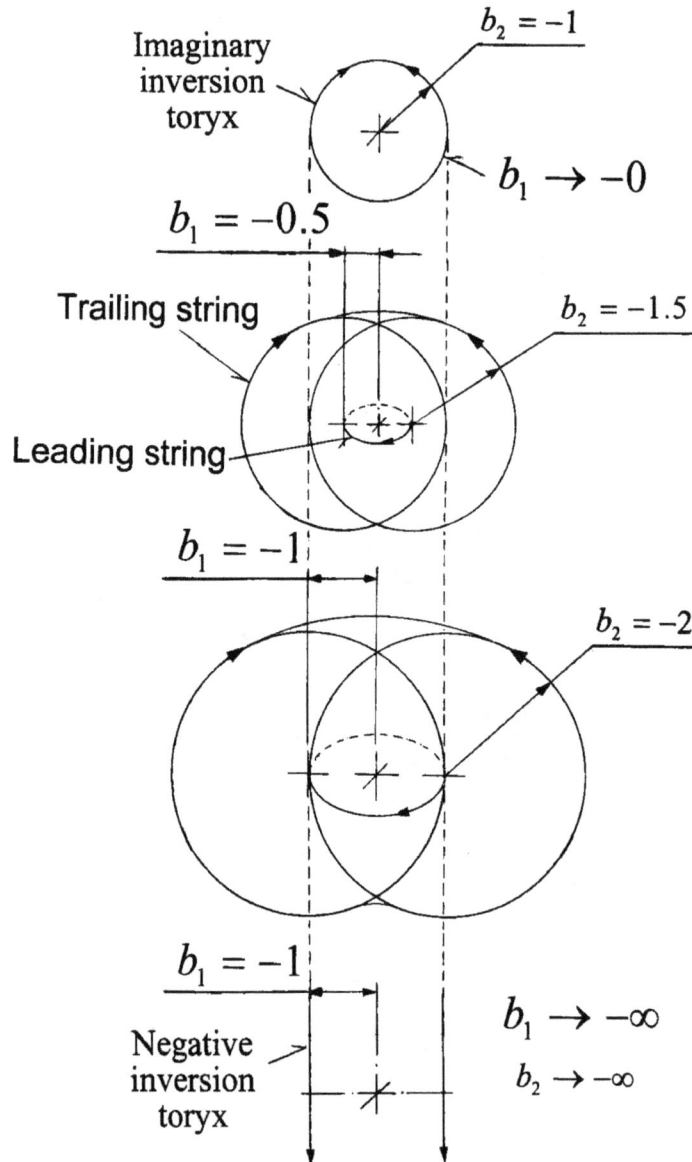

Range	φ_2	b	b_1	b_2	η_2	w_2	β_{2t}	β_{2r}
From	270^0	-1.0	-0	-1.0	$-i$	$+0i$	$-\infty i$	$+\infty$
To	360^0	$-\infty$	$-\infty$	$-\infty$	$-\infty i$	$+\infty i$	$-0i$	1.0

Figure 5.6. Transformations of imaginary negative toryces.

5.7 Summary of Toryx Transformations

Figure 5.7 summarizes transformation of toryces by showing together the metamorphoses of toryx spherical boundary, leading string and trailing string with the respective relative radii b, b_1 and b_2 as the steepness angle of trailing string φ_2 increases from 0^0 to 360^0.

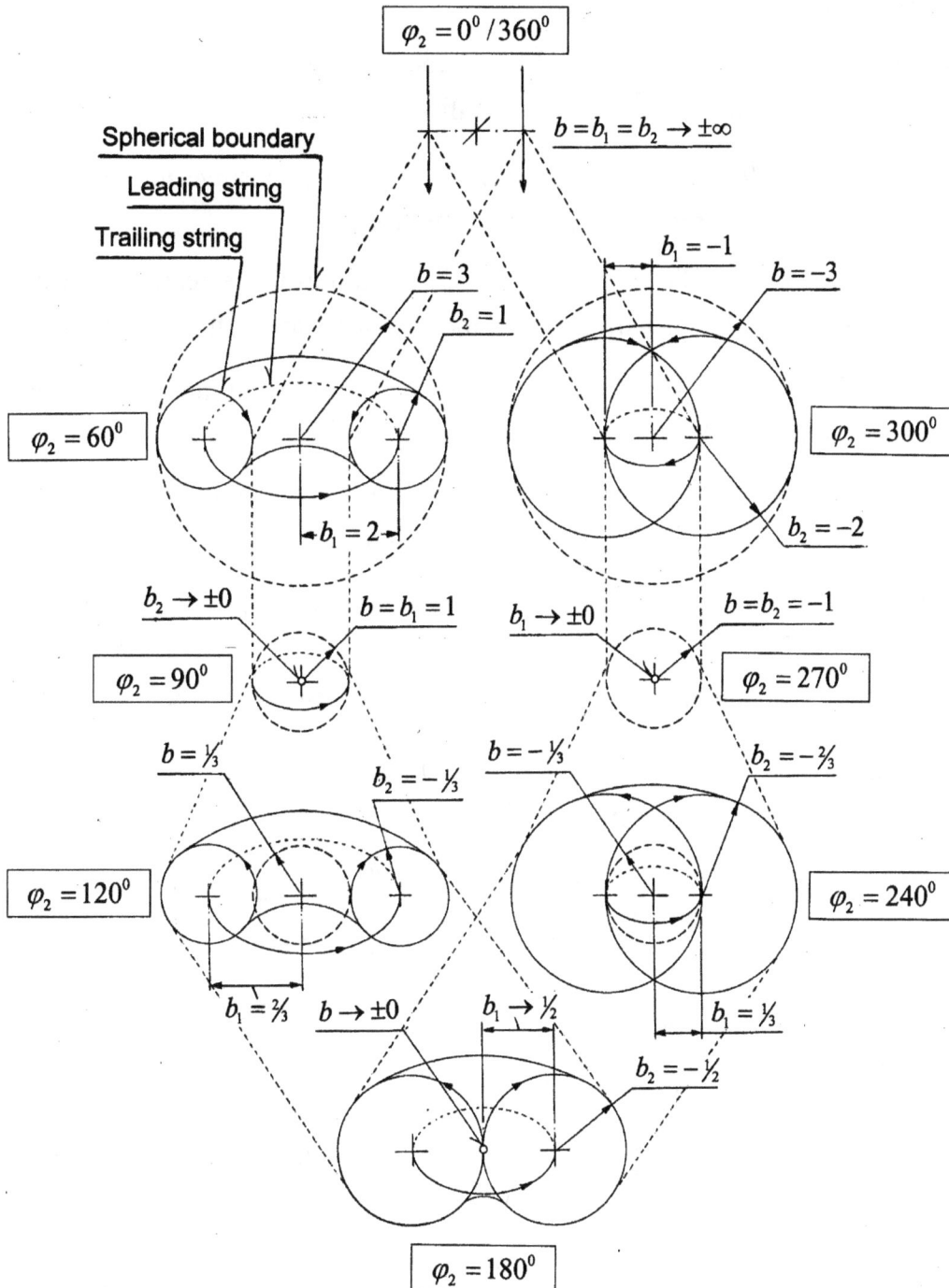

Figure 5.7. Transformations of toryx spherical boundary, leading string and trailing string as a function of the steepness angle of trailing string φ_2.

- When $\varphi_2 \to +0^0$, all three radii b, b_1 and b_2 approach positive infinity $(+\infty)$.

- Within the range of $\varphi_2 (+0^0 < \varphi_2 < 90^0)$, b, b_1 and b_2 decrease as φ_2 increases.

- When $\varphi_2 \to 90^0$, $b = b_1 = 1$ and b_2 approaches positive infinility $(+0)$; the *trailing string becomes inverted* and b_2 becomes negative.

- Within the range of $\varphi_2 (90^0 < \varphi_2 < 180^0)$, b_1 and b continue to decrease, while negative values of b_2 increase starting from negative infinility (-0).

- When $\varphi_2 \to 180^0$, $b_1 = \frac{1}{2}$, $b_2 = -\frac{1}{2}$ and b approaches positive infinility $(+0)$; the *toryx spherical boundary becomes inverted* and b becomes negative.

- Within the range of $\varphi_2 (180^0 < \varphi_2 < 270^0)$, b_1 continues to decrease, negative values of b_2 continue to increase, while negative values of b increase starting from negative infinility (-0).

- When $\varphi_2 \to 270^0$, $b = b_2 = -1$, while b_1 approaches positive infinility $(+0)$; the *leading string becomes inverted* and b_1 becomes negative.

- Within the range of $\varphi_2 (270^0 < \varphi_2 < 360^0)$, negative values of b_1 increase starting from negative infinility (-0), while negative values of b and b_2 continue to increase.

- When $\varphi_2 \to 360^0$, all three radii b, b_1 and b_2 approach negative infinity $(-\infty)$.

Table 5.7. Inversion points of toryx leading string, trailing strings and spherical boundary.

Toryx component	Relative radius	Steepness angle
Spherical boundary	$b \to \pm\infty$	$\varphi_2 \to 0^0/360^0$
	$b \to \pm 0$	$\varphi_2 \to 180^0$
Leading string	$b_1 \to \pm\infty$	$\varphi_2 \to 0^0/360^0$
	$b_1 \to \pm 0$	$\varphi_2 \to 270^0$
Trailing string	$b_2 \to \pm\infty$	$\varphi_2 \to 0^0/360^0$
	$b_2 \to \pm 0$	$\varphi_2 \to 90^0$

Table 5.7 summarizes the inversion steepness angles of trailing string and corresponding radii of toryx leading string, trailing string and spherical boundary.

5.8 Golden Toryces

Figure 5.8 and Table 2.4 show cross-sections and parameters of four golden toryces corresponding to the toryx golden polarization number $G = \pm1$.

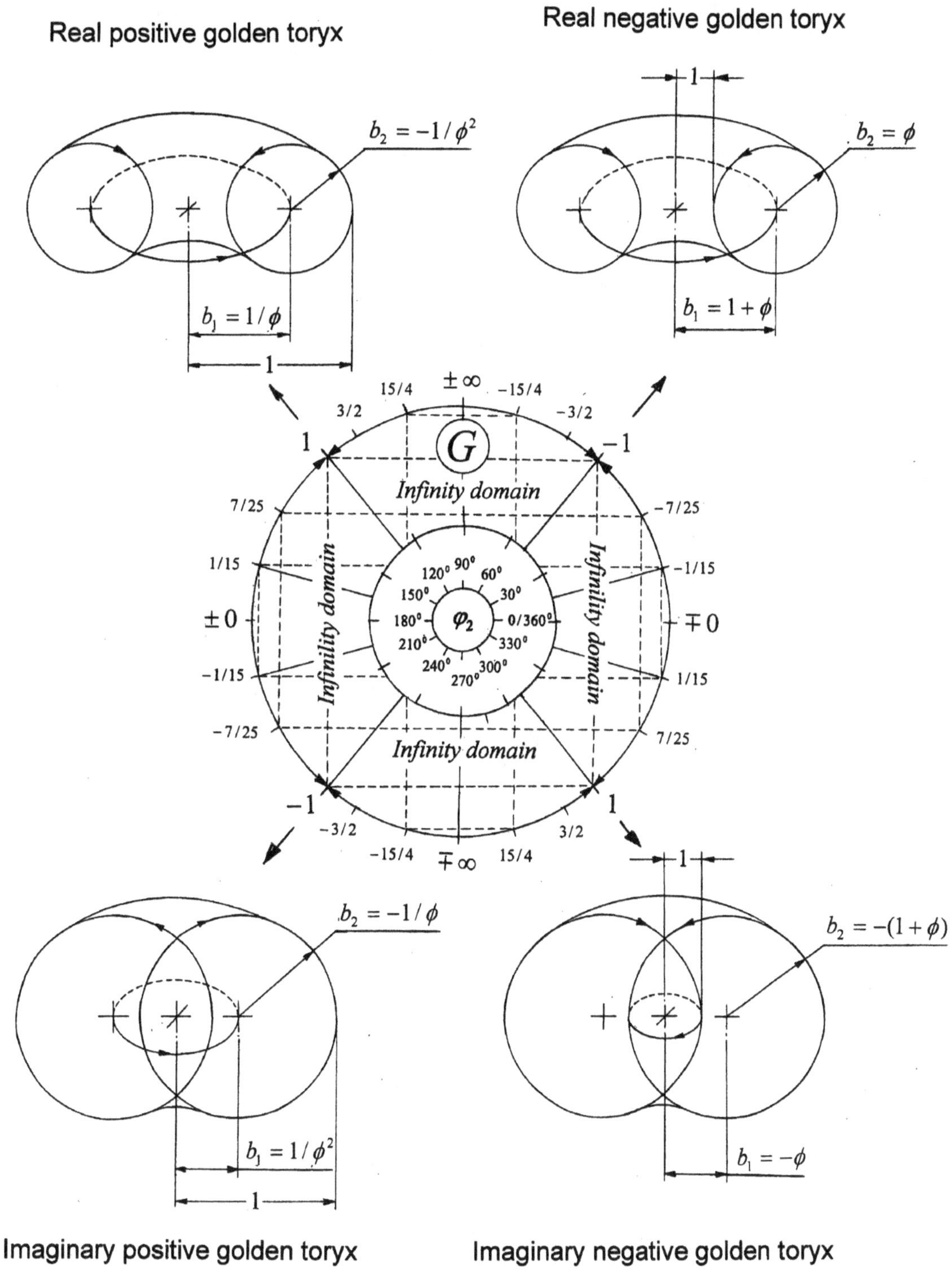

Figure 5.8 Golden toryces corresponding to the toryx golden polarization number $G = \pm1$.

Notes

PART 2

Applied
Mathematics
of a Toryx

6. QUANTUM STATES OF TORYCES

CONTENTS

6.1 Definition of Quantum States of Excited Toryces

Toryces change their dimensions in quantum steps by a so-called *excitation process*. As shown in Figure 6.1, during the excitation of a toryx the radius of toryx leading string r_1 increases, while its eye radius r_0 remains constant.

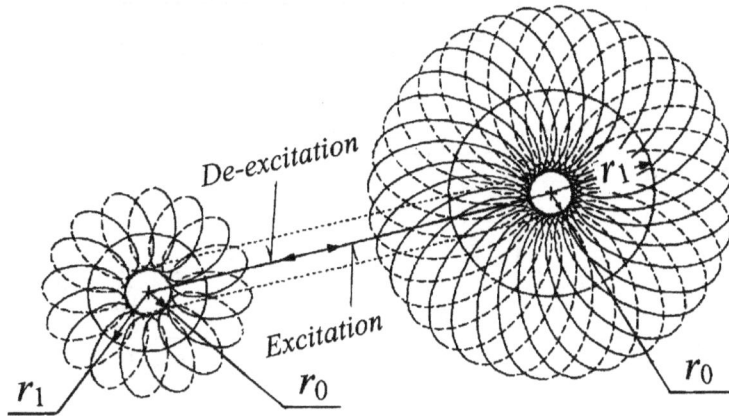

Figure 6.1. Excitation and de-excitation of toryces.

The derived quantization equations are based on three proposed limitations of degrees of freedom of real negative *lambda, harmonic* and *golden toryces* shown in Exhibit 6.1.

Exhibit 6.1. Limitations of degrees of freedom of excited toryces.

The relative radius of leading string of real negative toryx b_1 is equal to:		
Lambda toryx	**Harmonic toryx**	**Golden toryx**
$b_1 = z = 2(n\Lambda)^m$ (6.1-1)	$b_1 = z = 2 + n$ (6.1-2)	$b_1 = z = 2 + n/\phi$ (6.1-3)

In Exhibit 6.1:

z = toryx quantization parameter

$m \rightarrow 0, 1, 2, . . $, toryx exponential excitation quantum states

$n \rightarrow 0, 1, 2, . . $, toryx linear excitation quantum states

$\Lambda = 137, ...$, *toryx quantization constant*.

Notably, when $m \rightarrow 0$ and $n \rightarrow 0$, all three Eqs. (6.1-1) - (6.1-3) yield $z = 2$.

6.2 Quantization Equations for Excited Toryces

Table 6.2.1 shows quantization equations for relative radii of leading strings and spherical boundaries of excited lambda toryces.

Table 6.2.1. Quantization equations for relative radii of toryx leading strings and spherical boundaries of excited lambda toryces.

Lambda toryx	Relative radius of leading string	Relative radius of spherical boundary
$E^-_{m,n,q}$	$b^-_{1E} = z$ (6.2-1)	$b^-_E = 2z - 1$ (6.2-2)
$\breve{E}^-_{m,n,q}$	$\breve{b}^-_{1E} = 1 - z$ (6.2-3)	$\breve{b}^-_E = 1 - 2z$ (6.2-4)
$E^+_{m,n.q}$	$b^+_{1E} = \dfrac{z}{2z-1}$ (6.2-5)	$b^+_E = \dfrac{1}{2z-1}$ (6.2-6)
$\breve{E}^+_{m,n.q}$	$\breve{b}^+_{1E} = \dfrac{1-z}{1-2z}$ (6.2-7)	$\breve{b}^+_E = \dfrac{1}{1-2z}$ (6.2-8)
$A^-_{m,n,q}$	$b^-_{1A} = \dfrac{z}{z-1}$ (6.2-9)	$b^-_A = \dfrac{z+1}{z-1}$ (6.2-10)
$A^+_{m,n.q}$	$b^+_{1A} = \dfrac{z}{z+1}$ (6.2-11)	$b^+_A = \dfrac{z-1}{z+1}$ (6.2-12)
$\breve{A}^-_{m,n,q}$	$\breve{b}^-_{1A} = \dfrac{1}{1-z}$ (6.2-13)	$\breve{b}^-_A = \dfrac{1+z}{1-z}$ (6.2-14)
$\breve{A}^+_{m,n.q}$	$\breve{b}^+_{1A} = \dfrac{1}{1+z}$ (6.2-15)	$\breve{b}^+_A = \dfrac{1-z}{1+z}$ (6.2-16)

The first subscripts in the symbols of toryces are:

    For lambda toryces: m is standing for the toryx exponential excitation quantum state

    For harmonic toryces: H is standing for harmonic toryces with $m = 0$

    For golden toryces: G is standing for golden toryces with $m = 0$.

The second subscript indicates the toryx linear excitation quantum state n.
The third subscript indicates the oscillation quantum states q to be described in Section 6.4.
The superscript indicates the sign of toryx vorticities V and, in some particular cases, their values.

Table 6.2.2 shows quantization equations for relative radii of leading strings and spherical boundaries of excited harmonic toryces. In the harmonic toryces, the frequencies of their trailing strings are related to one another by simple harmonic ratios, explaining their names.

Table 6.2.2. Quantization equations for relative radii of toryx leading strings and spherical boundaries of excited harmonic toryces.

Harmonic toryx	Relative radius of leading string		Relative radius of spherical boundary	
$E_{H,n,q}^{-}$	$b_{1E}^{-} = 2+n$	(6.2-17)	$b_{E}^{-} = 3+2n$	(6.2-18)
$\breve{E}_{H,n,q}^{-}$	$\breve{b}_{1E}^{-} = -(1+n)$	(6.2-19)	$\breve{b}_{E}^{-} = -(3+2n)$	(6.2-20)
$E_{H,n.q}^{+}$	$b_{1E}^{+} = \dfrac{2+n}{3+2n}$	(6.2-21)	$b_{E}^{+} = \dfrac{1}{3+2n}$	(6.2-22)
$\breve{E}_{H,n.q}^{+}$	$\breve{b}_{1E}^{+} = \dfrac{1+n}{3+2n}$	(6.2-23)	$\breve{b}_{E}^{+} = -\dfrac{1}{3+2n}$	(6.2-24)
$A_{H,n.q}^{-}$	$b_{1A}^{-} = \dfrac{2+n}{1+n}$	(6.2-25)	$b_{A}^{-} = \dfrac{3+n}{1+n}$	(6.2-26)
$A_{H,n.q}^{+}$	$b_{1A}^{+} = \dfrac{2+n}{3+n}$	(6.2-27)	$b_{A}^{+} = \dfrac{1+n}{3+n}$	(6.2-28)
$\breve{A}_{H,n.q}^{-}$	$\breve{b}_{1A}^{-} = -\dfrac{1}{1+n}$	(6.2-29)	$\breve{b}_{A}^{-} = -\dfrac{3+n}{1+n}$	(6.2-30)
$\breve{A}_{H,n.q}^{+}$	$\breve{b}_{1A}^{+} = \dfrac{1}{3+n}$	(6.2-31)	$\breve{b}_{A}^{+} = -\dfrac{1+n}{3+n}$	(6.2-32)

For the particular cases when $z = 0$ and $n = 0$, both sets of equations for eshown in Tables 6.2.1 and 6.2.2 yield the quantum states of harmonic toryces located at the borderlines between polarized toryces, so the following adjacent harmonic toryces have the same structures and properties:

$$A_{H,n.q}^{-} = E_{H,n.q}^{-}; \ A_{H,n.q}^{+} = E_{H,n.q}^{+}; \ \breve{A}_{H,n.q}^{-} = \breve{E}_{H,n.q}^{-}; \ \breve{A}_{H,n.q}^{+} = \breve{E}_{H,n.q}^{+}$$

6.3 Quantization Equations for Resonant Excited Lambda Toryces

In the resonant negative toryx $\breve{E}r^-_{m,n,q}$, the frequency of its trailing string has the same amplitudes but opposite signs in respect to the basic toryx $E^-_{m,n,q}$. The quantization equations for the resonant negative toryx $\breve{E}r^-_{m,n,q}$ and the respective positive toryx $\breve{E}r^+_{m,n,q}$ are shown in Table 6.3.

Table 6.3. Quantization equations for the relative radii of leading strings and spherical boundary of basic and resonant excited lambda toryces.

Toryces	Relative radius of leading string	Relative radius of spherical boundary
Resonant negative toryx $\breve{E}r^-_{m,n,q}$	$\breve{b}^-_{1Er} = -z$ (6.3-1)	$\breve{b}^-_{Er} = -2z - 1$ (6.3-2)
Resonant positive toryx $\breve{E}r^+_{m,n,q}$	$\breve{b}^+_{1Er} = \dfrac{z}{2z+1}$ (6.3-3)	$\breve{b}^+_{Er} = -\dfrac{1}{2z+1}$ (6.3-4)

6.4 Quantization Equations for Oscillated Toryces

During oscillation of a toryx, its radius of leading string r_1 and its eye radius r_0 change proportionally as shown in Fig. 6.4.1 in comparison with the toryx excitation during which the toryx eye radius r_0 remains constant, while the radius of its leading string r_1 increases.

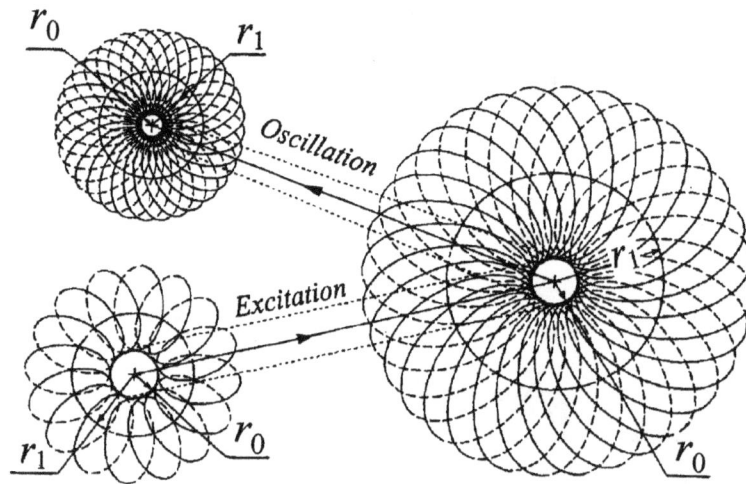

Figure 6.4.1. Oscillation of a toryx versus excitation of a toryx.

The oscillation of a toryx is a function of the ***toryx oscillation factor*** Q_q calculated based on the proposed limitation of its degree of freedom shown in Exhibit 6.4.

Exhibit 6.4. Limitations of degrees of freedom of oscillated toryces.

The toryx oscillation factor Q_q is equal to:

$$Q_0 = 1; \quad Q_q = 3\left(\frac{\Lambda}{2(q-1)}\right)^{q-1} \tag{6.4-1}$$

where

$q = 0, 1, 2, \ldots,$ ***toryx oscillation quantum states***.

The values of Q_q calculated from Eq. (6.4-1) are shown in Figure 6.4.2. At $q = q_m$ and $\Lambda = 137$, the toryx oscillation factor Q_q reaches its maximum value Q_{qm} defined by the equations:

$$q_m = 1 + \frac{\Lambda}{2e} = 26.199742$$

$$Q_{qm} = 3e^{\Lambda/2e} = 2.637728 \times 10^{11} \tag{6.4-2}$$

When $q > q_m$, the toryx oscillation factor Q_q sharply decreases and at $q = 70.589986$ its magnitude reduces to 1. After that as q continues to increase and approaches infinity, Q_q decreases and approaches infinility.

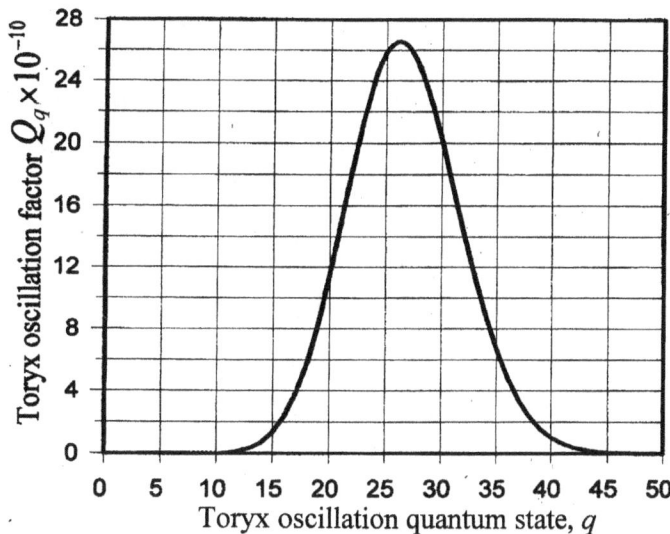

Figure 6.4.2. Toryx oscillation factor Q_q as a function of the toryx oscillation state q.

Figure 6.4.3 shows a plot of the natural logarithm of the toryx oscillation factor $\ln Q_q$ and the toryx oscillation factor Q_q as a function of the toryx oscillation quantum states q calculated from Eq. (6.4-1).

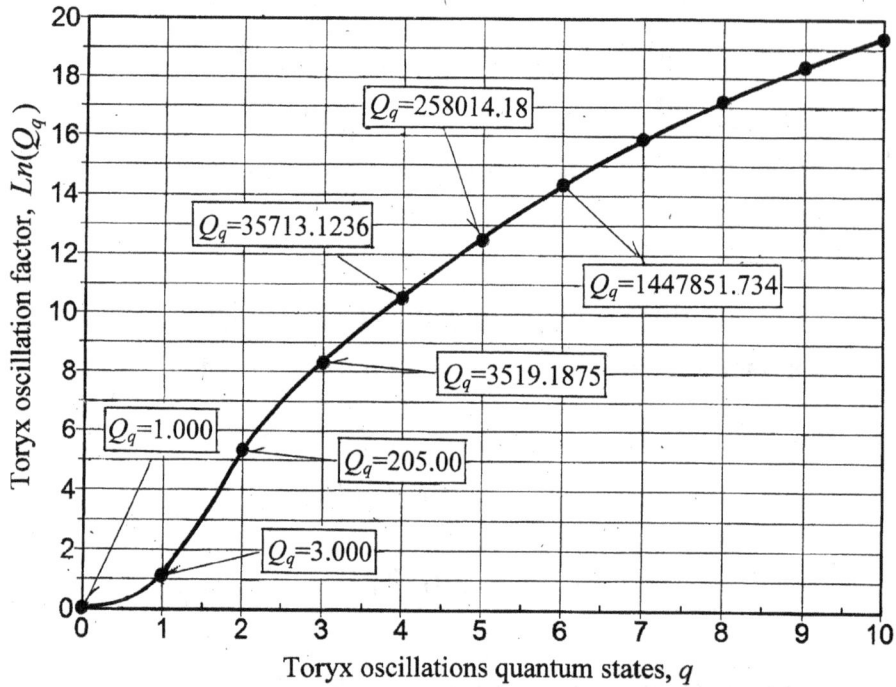

Figure 6.4.3. Toryx oscillation factor Q_q as a function of the toryx oscillation quantum states q.

6.5 Toryx Fine Structure

As shown in Figs. 6.5.1 and 6.5.2, fine structures of negative real and imaginary toryces are formed by *standing waves* oscillating inside *toryx fine-structure spherical boundary* with the radius r_3. Each standing wave is made up of a leading string with the radius $r_4 \rightarrow \infty$ and a double-helical trailing with the radius r_5.

The radius of toryx fine-structure spherical boundary r_3 is equal to:

$$r_3 = \sqrt{r_1^2 - r_2^2} \tag{6.5-1}$$

The relative radius of toryx fine-structure spherical boundary b_3 is equal to:

$$b_3 = \eta_2 = \sqrt{2b_1 - 1} \tag{6.5-2}$$

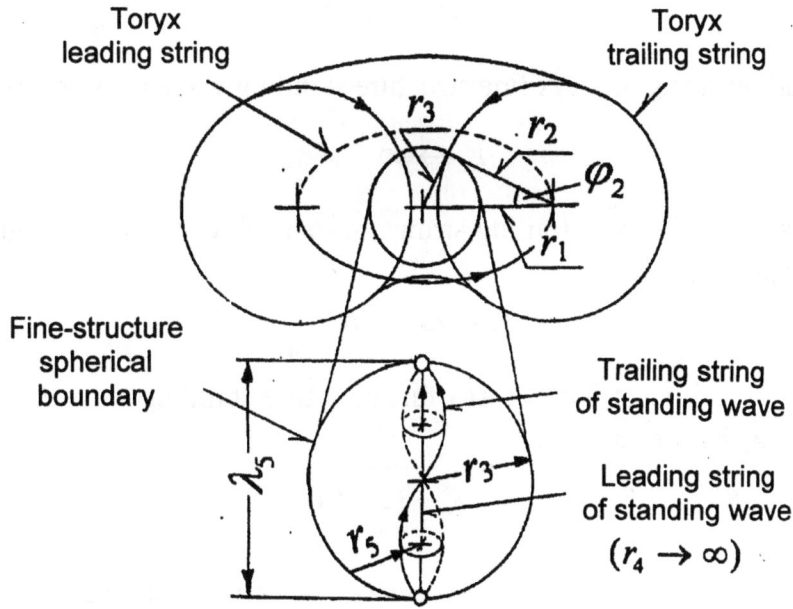

Figure 6.5.1. Fine structure of a real negative toryx.

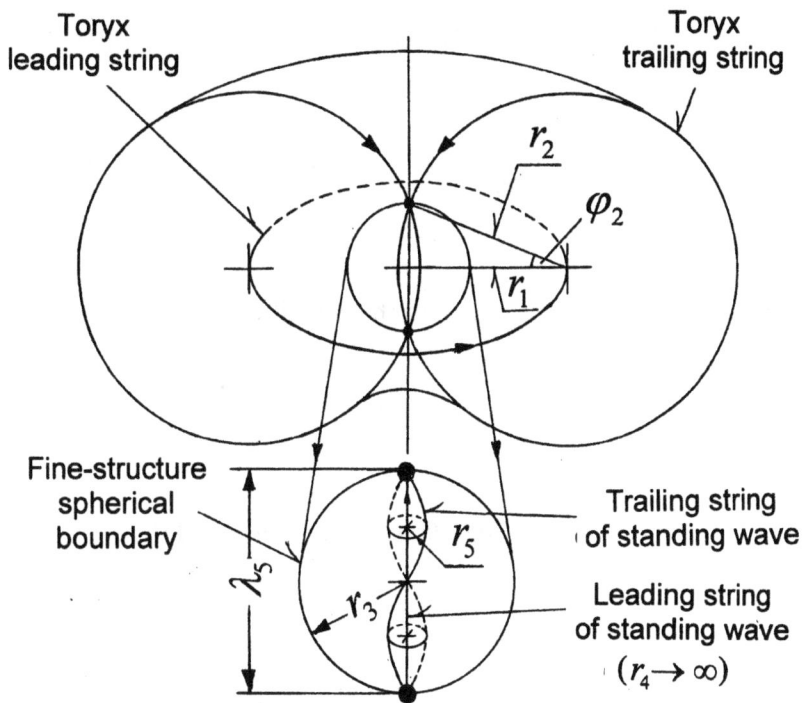

Figure 6.5.2. Fine structure of an imaginary negative toryx.

The perimeter $2\pi r_3$ of toryx fine-structure spherical boundary is equal to the wavelength λ_2 of toryx trailing string:

$$2\pi r_3 = \lambda_2 \qquad\qquad (6.5\text{-}3)$$

The radius of leading string of toryx fine-structure standing wave r_4 approaches infinity:

$$r_4 \to \infty \qquad\qquad (6.5\text{-}4)$$

The wavelengths of leading and trailing strings of toryx fine-structure standing wave, λ_4 and λ_5, are the same and equal to:

$$\lambda_4 = \lambda_5 = 2r_3 \qquad\qquad (6.5\text{-}5)$$

The spiral length of trailing string of toryx fine-structure standing wave L_5 is related to its radius r_5 and wavelength λ_5 by the equation:

$$L_5 = \sqrt{\lambda_5^2 + (2\pi r_5)^2} \qquad\qquad (6.5\text{-}6)$$

To define the radius r_5 of trailing string of toryx fine-structure standing wave, it is necessary to introduce the limitation of its degree of freedom shown in Exhibit 6.5.

Exhibit 6.5. Limitation of a degree of freedom of toryx fine-structure standing wave.

> The aspect ratio of fine-structure standing wave $\lambda_5/2r_5$ is equal to the toryx quantization constant Λ:
>
> $$\frac{\lambda_5}{2r_5} = \Lambda = const. \qquad\qquad (6.5\text{-}7)$$

From Eqs. (6.5-6) and (6.5-7), the spiral length of trailing string of toryx fine-structure standing wave L_5 is equal to:

$$L_5 = \lambda_5 \frac{\alpha_s^{-1}}{\Lambda} \qquad\qquad (6.5\text{-}8)$$

where α_s^{-1} is the ***spacetime inverse fine-structure constant*** related to the spacetime quantization constant Λ by the equation:

$$\alpha_s^{-1} = \sqrt{\Lambda^2 + \pi^2} \qquad\qquad (6.5\text{-}9)$$

Notably, Eq. (6.5-9) is similar to the equation proposed by T.J. Burger as an approximation of the inverse fine structure constant. For the case when $\Lambda = 137$, this equation yields the value of $\alpha_s^{-1} = 137.036015720$. The relative difference between this values and the experimental value of the inverse fine structure constant $\alpha^{-1} = 137.035999074$ provided by 2011 CODATA is about one part per ten million.

7. FORMATION OF ELEMENTARY MATTER PARTICLES

7.1 Relationships between Physical & Spacetime Properties of Toryces

The following physical constants are used for describing toryx physical parameters:

c = velocity of light
e = elementary charge
l_p = Planck length
m_e = electron mass
m_p = proton mass
r_e = classical electron radius
μ_B = Bohr magneton
μ_N = nuclear magneton
α_s = spacetime fine structure constant
ε_0 = electric constant.

Derivation of relationships between toryx physical and spacetime parameters are based on two postulates shown in Exhibit 7.1.

Exhibit 7.1. Basic relationships between toryx physical and spacetime parameters.

- The relative toryx charge e_t / e_0 is equal to the toryx vorticity V:

$$\frac{e_t}{e} = V \qquad (7.1\text{-}1)$$

- The relative toryx gravitational mass m_g / m_e is proportional to the absolute value of the toryx vorticity $|V|$:

$$\frac{m_g}{m_e} = Q_q |V| \qquad (7.1\text{-}2)$$

Tables 7.1.1 and 7.1.2 show derived relationships between toryx physical and spacetime parameters.

Table 7.1.1. Relationships between toryx relative physical and spacetime parameters.

Toryx relative parameters	Equations	
Charge	$$\frac{e_t}{e} = -\frac{r_2}{r_1} = -\frac{b_1-1}{b_1} = V$$	(7.1-3)
Gravitational mass	$$\frac{m_g}{m_e} = Q_q\left|\frac{r_2}{r_1}\right| = Q_q\left|\frac{b_1-1}{b_1}\right| = Q_q|V|$$	(7.1-4)
Inertial mass	$$\frac{m_{ti}}{m_e} = Q_q\frac{2r_2}{r} = Q_q\frac{2(b_1-1)}{2b_1-1} = -\frac{2Q_qV}{\delta_2 R^2}$$	(7.1-5)
Magnetic moment	Real toryces: $\dfrac{\mu_t}{\mu_B} = \pm\alpha_s\dfrac{(b_1-1)\sqrt{2b_1-1}}{2Q_qb_1} = \pm\alpha_s\dfrac{VR}{2Q_q}$ Imaginary toryces: $\dfrac{\breve{\mu}_t}{\mu_B} = \pm\alpha_s i\dfrac{(\breve{b}_1-1)\sqrt{2\breve{b}_1-1}}{2Q_q\breve{b}_1} = \pm\alpha_s i\dfrac{VR}{2Q_q}$	(7.1-6)
Matter energy	$$\frac{E_m}{m_ec^2}V = Q_q\frac{b_1-1}{b_1} = -Q_qV$$	(7.1-7)
Field energy	$$\frac{E_f}{m_ec^2} = \frac{Q_q}{b_1} = Q_q\delta_2$$	(7.1-8)
Kinetic energy	$$\frac{K_t}{m_ec^2} = Q_q\frac{(b_1-1)}{b_1^2} = -Q_qV\delta_2$$	(7.1-9)
Potential energy	$$\frac{U_t}{m_ec^2} = -\frac{2Q_q(b_1-1)}{b_1^2} = 2Q_qV\delta_2$$	(7.1-10)
Total kinetic & potential energy	$$\frac{E_t}{m_ec^2} = -Q_q\frac{(b_1-1)}{b_1^2} = Q_qV\delta_2$$	(7.1-11)
Spacetime intensity	$$I = Q_qG = -Q_q\frac{2b_1-1}{(b_1-1)b_1} = Q_q\frac{(R\delta_2)^2}{V}$$	(7.1-12)
Density	$$\rho_t\frac{2\pi^2r_0^3}{m_e} = \frac{Q_q}{b_1(b_1-1)^2}\left|\frac{b_1-1}{b_1}\right| = \frac{Q_q\delta_2^3}{|V|}$$	(7.1-13)
Modulus of elasticity	$$B_t\frac{2\pi^2r_0^3}{m_ec^2} = Q_q\left|\frac{b_1-1}{b_1}\right|\frac{2b_1-1}{b_1^3(b_1-1)^2} = \frac{Q_qR^2\delta_2^5}{|V|}$$	(7.1-14)

Table 7.1.2. Relationships between toryx constant spacetime and physical parameters.

Toryx spacetime parameters	Equations	
Eye radius	$$r_0 = \frac{r_e}{2} = \frac{e^2}{8\pi\varepsilon_0 m_e c^2}$$	(7.1-15)
Base frequency	$$f_0 = \frac{c}{2\pi r_0} = \frac{4\varepsilon_0 m_e c^3}{e^2}$$	(7.1-16)
Base period	$$T_0 = \frac{2\pi r_0}{c} = \frac{e^2}{4\varepsilon_0 m_e c^3}$$	(7.1-17)

Toryx relative magnetic moment – Figure 7.1.1 shows a plot of Eq. (7.1-6) for the toryx relative magnetic moment μ_t / μ_B expressed in respect to the Bohr magneton μ_B given by the equation:

$$\mu_B = \frac{e^3}{8\pi\alpha_s \varepsilon_0 m_e c}$$

(7.1-18)

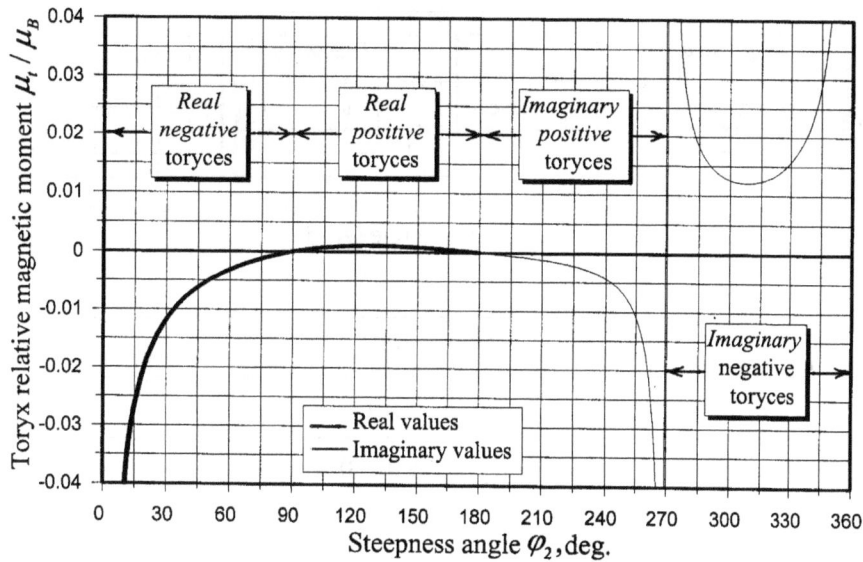

φ_2	$360^0 / 0^0$	90^0	180^0	270^0
μ_t / μ_B	$+\infty i / -\infty$	$-0 / +0$	$+0 / -0i$	$-\infty i / +\infty i$

Figure 7.1.1. Toryx relative magnetic moment in respect to the Bohr magneton μ_t / μ_B as a function of the steepness angle of trailing string φ_2 when $Q_q = 1$.

The toryx relative magnetic moment in respect to the nuclear magneton μ_t / μ_N is related to the toryx relative magnetic moment in respect to the Bohr magneton μ_t / μ_B by the equation:

$$\frac{\mu_t}{\mu_N} = \frac{\mu_t}{\mu_B} \frac{m_p}{m_e} \qquad (7.1\text{-}19)$$

Toryx field & matter energy - Figure 7.1.2 shows plots of Eqs. (7.1-7) and (7.1-8) for the toryx relative field energy $E_f / m_e c^2$ and matter energy $E_m / m_e c^2$ as functions of the steepness angle of trailing string φ_2 when $Q_q = 1$.

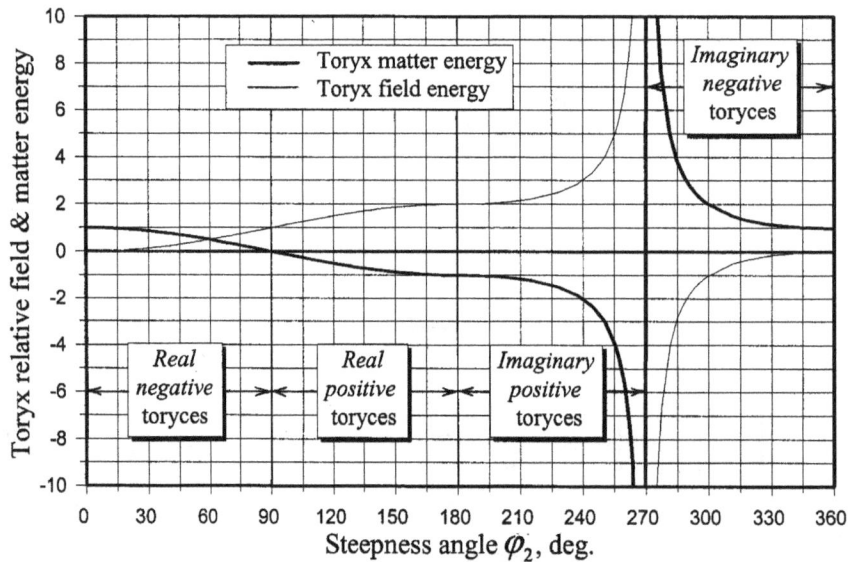

φ_2	$360^0 / 0^0$	90^0	180^0	270^0
$E_f / m_e c^2$	$-0/+0$	$+1$	$+2$	$+\infty / -\infty$
$E_m / m_e c^2$	$+1.0$	$+0/-0$	-1.0	$-\infty / +\infty$

Figure 7.1.2. Toryx relative field energy $E_f / m_e c^2$ and matter energy $E_m / m_e c^2$ as functions of the steepness angle of trailing string φ_2 when $Q_q = 1$.

Toryx density – Figure 7.1.3 shows a plot of Eq. (7.1-13) for the toryx relative density ρ_{tr}. It is derived based on the assumption equal to the ratio of the toryx gravitational mass m_{tg} to the volume occupied by its trailing string.

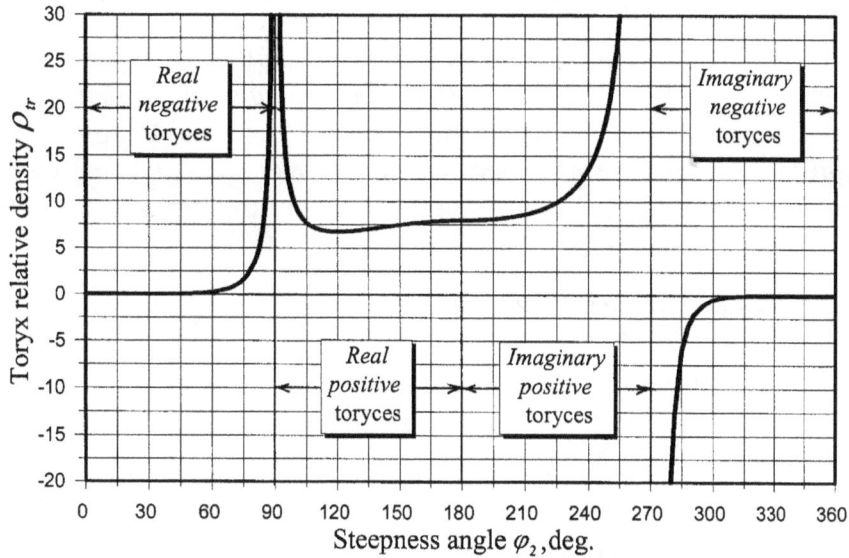

φ_2	$360^0/0^0$	90^0	180^0	270^0
ρ_{tr}	$-0/+0$	$+\infty$	8.0	$+\infty/-\infty$

Figure 7.1.3. Toryx relative density ρ_{tr} as a function of the steepness angle of trailing string φ_2 when $Q_q = 1$.

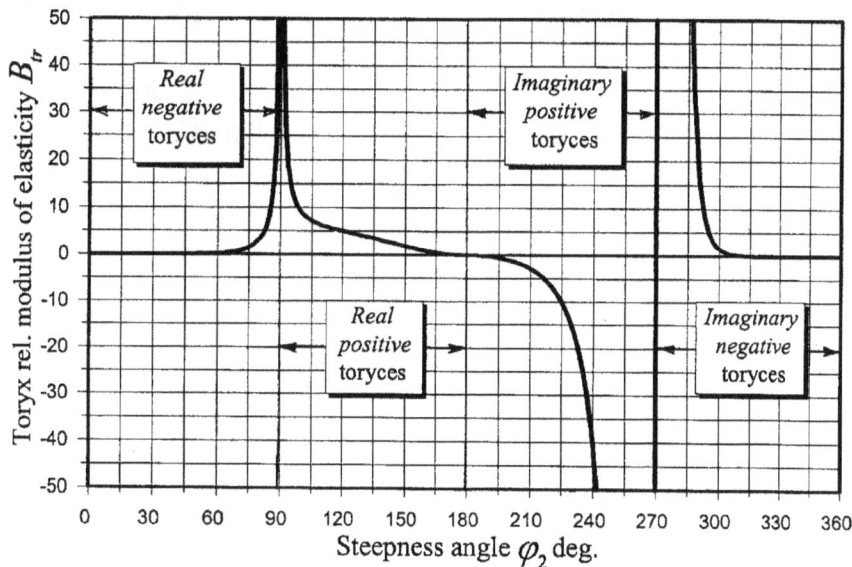

φ_2	$360^0/0^0$	90^0	180^0	270^0
B_{tr}	0	$+\infty$	$+0/-0$	$-\infty/+\infty$

Figure 7.1.4 Toryx relative bulk modulus of elasticity B_{tr} as a function of the steepness angle of trailing string φ_2.

Toryx modulus of elasticity – Figure 7.1.4 shows a plot of Eq. (7.1-14) for the toryx relative modulus of elasticity B_{tr}. In classical mechanics, elastic properties, or rigidity, of compressible media contained in a certain volume are defined by the ***bulk modulus of elasticity***. It is equal to the ratio of change in pressure inside the media to the resulting fractional change in the volume of the media. The compression waves are assumed to behave like sound waves, traveling through the toryx trailing string with velocity equal to the spiral velocity of the toryx leading string V_1.

Toryx relativistic equations are shown in Table 7.1.3.

- Both masses and charges of toryces are dependent on velocities of their leading strings.
- In real toryces the absolute values of their relative charges and masses decrease with an increase of the relative spiral velocities of their leading strings β_1.
- In imaginary toryces the absolute values of their relative charges and masses increase as the relative spiral velocities of their leading string β_1 increase.

Table 7.1.3. Toryx relativistic equations.

Relative charge	$\dfrac{e_t}{e} = \sqrt{1 - \beta_1^2}$	(7.1-20)
Relative gravitational mass	$\dfrac{m_{tg}}{m_e} = Q_q\left|\sqrt{1 - \beta_1^2}\right|$	(7.1-21)

7.2 Formation of Virtual Toryces

According to the proposed theory, the toryx spacetime limits correspond to the Planck length l_p below which the toryx spacetime properties are no longer defined by the toryx spacetime postulates, but by the proposed ***toryx uncertainty principle*** presented in Exhibit 7.2.

Exhibit 7.2. Toryx uncertainty principle.

A virtual toryx can be produced spontaneously in quantum vacuum and exist for the period equal to the relative period of its leading string t_1 if an absolute product of the toryx vorticity V and the period t_1 is equal or less than $1/(Q_q\pi\alpha_s)$ as given by the equation:

$$\text{Real toryces: } |Vt_1| = \frac{b_1(b_1-1)}{\sqrt{2b_1-1}} \leq \frac{1}{Q_q\pi\alpha_s} \tag{7.2-1}$$

$$\text{Imaginary toryces: } |Vt_1| = \frac{b_1(b_1-1)i}{\sqrt{2b_1-1}} \leq \frac{1}{Q_q\pi\alpha_s} \tag{7.2-2}$$

For the non-oscillated harmonic toryces $(Q_q = 1)$, we obtain from Eqs. (7.2-1) and (7.2-2):

$$0.5000164241 < b_1 < 16.1229570320 \qquad (7.2\text{-}3)$$

$$0.4999835759 > b_1 > -15.1229570320 \qquad (7.2\text{-}4)$$

Figure 7.2 shows ranges of the toryx relative radii of spherical boundaries b within which the spacetime ceases to exist and replaced by spontaneously appearing *virtual toryces* of quantum vacuum forming short-lived *virtual particles*.

The relative spacetime length b_s corresponding to the limits of spacetime is equal to:

$$b_s = \frac{l_p}{2\pi r_0} \qquad (7.2\text{-}5)$$

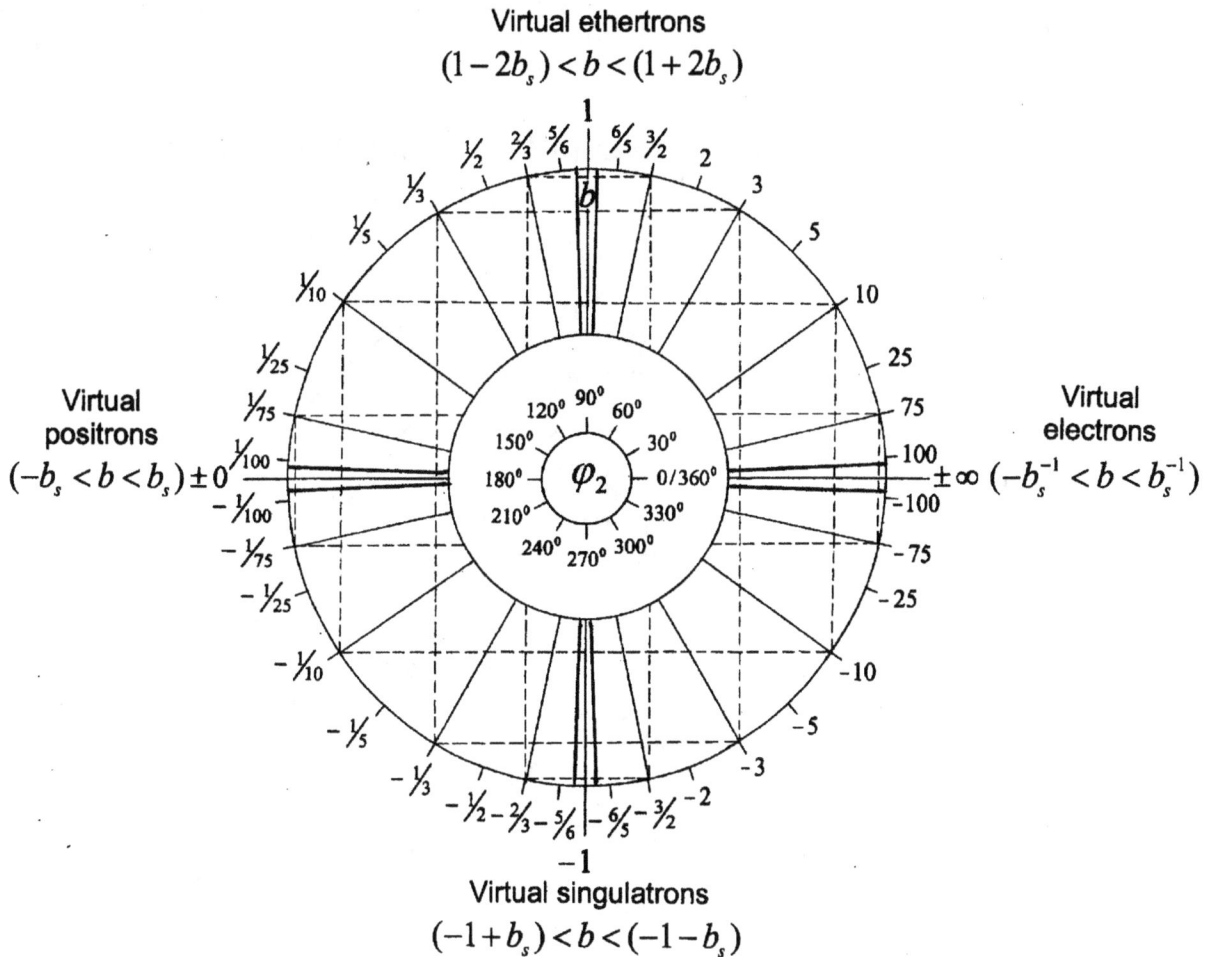

Figure 7.2. Ranges of the toryx relative radii of spherical boundaries b.

Table 7.2 shows the ranges of the toryx relative radii of spherical boundaries b, leading string b_1 and trailing string b_2 of electrons, positrons, ethertrons and singulatrons within which the space-time ceases to exist and replaced by the virtual particles of quantum vacuum.

Table 7.2. Ranges of toryx parameters within which spacetime ceases to exist
and replaced by quantum vacuum.

Virtual particles	Leading string	Trailing string	Spherical boundary
Electron	$\dfrac{1-b_s^{-1}}{2} < b_1 < \dfrac{1+b_s^{-1}}{2}$	$-\dfrac{1+b_s^{-1}}{2} < b_2 < -\dfrac{1-b_s^{-1}}{2}$	$-b_s^{-1} < b < b_s^{-1}$
Positron	$\dfrac{1-b_s}{2} < b_1 < \dfrac{1+b_s}{2}$	$-\dfrac{1+b_s}{2} < b_2 < -\dfrac{1-b_s}{2}$	$-b_s < b < b_s$
Ethertron	$1-b_s < b_1 < 1+b_s$	$-b_s < b_2 < b_s$	$1-2b_s < b < 1+2b_s$
Singulatron	$-b_s < b_1 < b_s$	$-1-b_s < b_2 < -1+b_s$	$-1-2b_s < b < -1+2b_s$

7.3   From Quantum Vacuum to Elementary Matter Particles

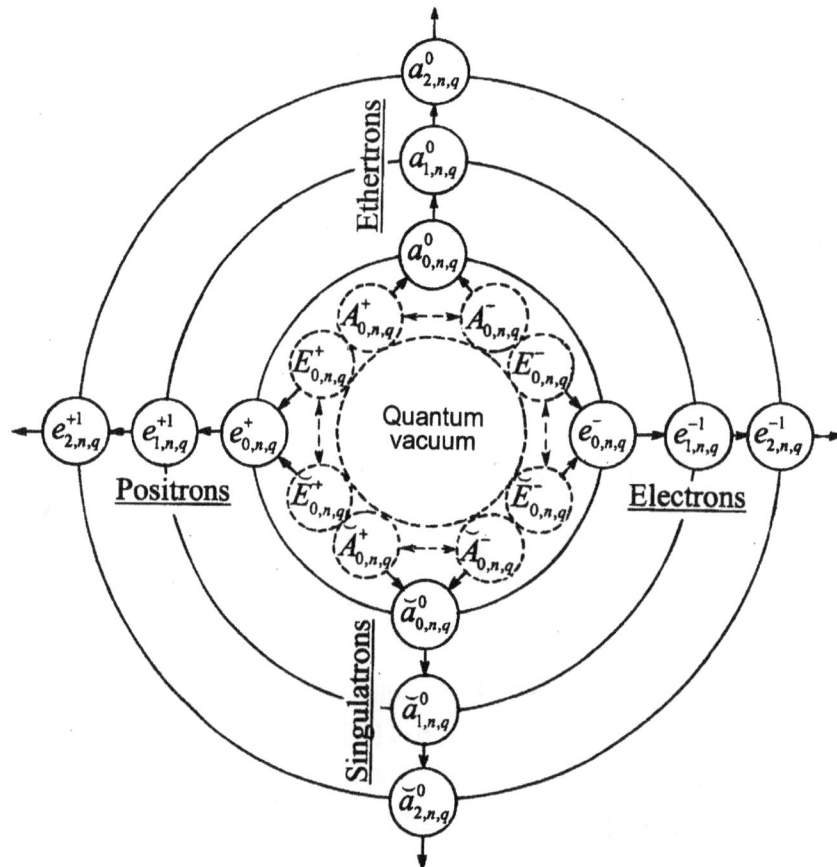

Figure 7.3.1. Formation of elementary particles (trons) from quantum vacuum.

Four basic elementary particles (trons) are formed from quantum vacuum by the unification of adjacent toryces: *electrons, positrons, ethertrons* and *singulatrons* as shown in Fig. 7.3.1.

The elementary particles (trons) are created in four steps:

1. Formation of *virtual toryces* from *quantum vacuum*
2. Formation of *harmonic toryces* from virtual toryces
3. Formation of *harmonic elementary particles* by unification
   of polarized harmonic toryces
4. Formation of *excited and oscillated elementary particles*
   from harmonic toryces.

The trons are divided into two kinds: *self-polarized trons and mutually-polarized trons*.

Figure 7.3.2 and Table 7.3.1 show general presentations of formation of self-polarized harmonic trons.

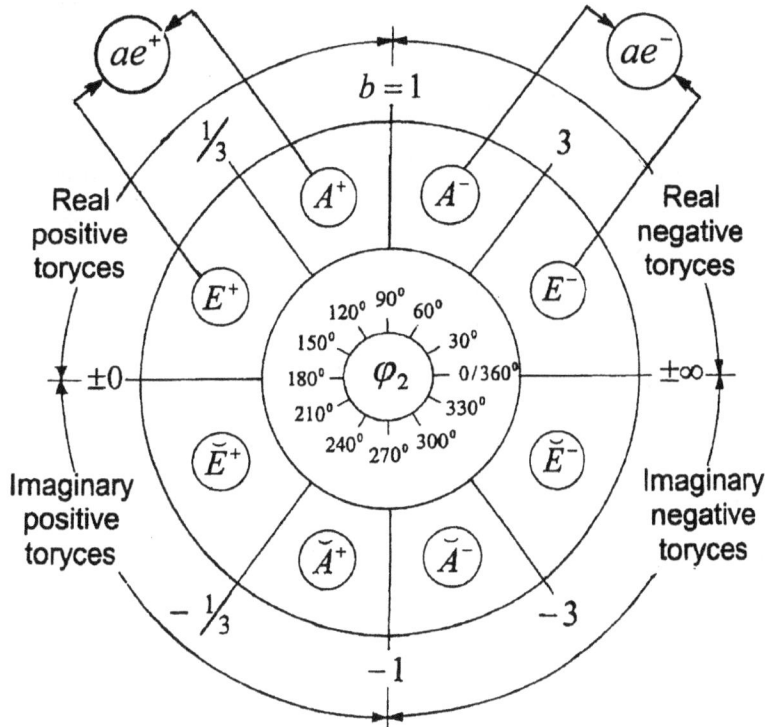

Figure 7.3.2. Formation of self-polarized harmonic trons.

Table 7.3.1. Constituent toryces of self-polarized harmonic trons.

Trons	Constituent toryces	Trons	Constituent Toryces
$ae_{0,n,q}^{-1}$	$A_{H,n,q}^{-1/2} + E_{H,n,q}^{-1/2}$	$ae_{H,n,q}^{+1}$	$A_{H,n,q}^{+1/2} + E_{H,n,q}^{+1/2}$

Figure 7.3.3 and Table 7.3.2 show general presentations of formation of four mutually-polarized excited lambda trons.

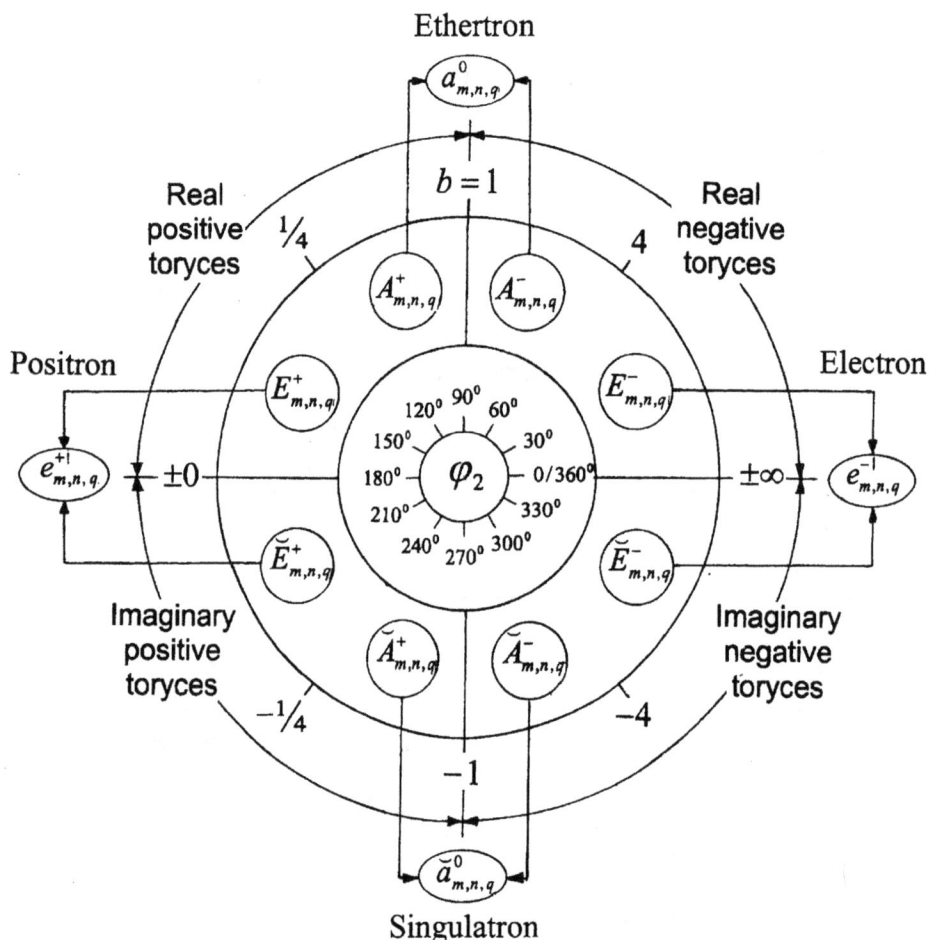

Figure 7.3.3. Formation of mutually-polarized excited lambda trons.

Table 7.3.2. Constituent toryces of mutually-polarized excited lambda trons.

Tron	Constituent toryces	Tron	Constituent toryces
Electron $e_{m,n,q}^{-1}$	$E_{m,n,q}^{-} + \breve{E}_{m,n,q}^{-}$	Ethertron $a_{m,n,q}^{0}$	$A_{m,n,q}^{-} + A_{m,n,q}^{+}$
Positron $e_{m,n,q}^{+1}$	$E_{m,n,q}^{+} + \breve{E}_{m,n,q}^{+}$	Singulatron $\breve{a}_{m,n,q}^{0}$	$\breve{A}_{m,n,q}^{-} + \breve{A}_{m,n,q}^{+}$

7.4 Rules of Formation of Stable Elementary Matter Particles

Trons sustain their existence by a periodic absorption and release of spacetime. Imaginary toryces are responsible for absorption of spacetime, while real toryces are responsible for release of spacetime. Stability of reality-polarized trons is governed by the proposed *tron polarization conservation law* (see Exhibit 7.4).

Exhibit 7.4. Tron polarization conservation law.

The *toryx polarization* P is given by the equation:

$$P = Q_q VR^2 = -\frac{Q_q(b_1 - 1)(2b_1 - 1)}{b_1} \qquad (7.4\text{-}1)$$

The *polarization* P_t of a stable tron made up of N real toryces with the polarization P and $\breve{N}$ imaginary toryces with the polarization $\breve{P}$ is infinitesimally small:

$$P_t = (PN + \breve{P}\breve{N}) \to \pm 0 \qquad (7.4\text{-}2)$$

Consequently, in a stable tron made up of N real and $\breve{N}$ imaginary toryces, the **stable tron reality ratio** T is defined by the equation:

$$T = \frac{N}{\breve{N}} = \left(\frac{b_1}{b_1 - 1}\right)^2 = \left(\frac{b+1}{b-1}\right)^2 \qquad (7.4\text{-}3)$$

where b_1 and b are respectively relative radii of toryx leading string and spherical boundary.

The following additional rules are applied to the calculations of parameters of mutually-polarized and self-polarized trons:

- Mutually-polarized trons are made up of either vorticity-polarized or reality-polarized toryces, coexisting **alternatively** inside their trons. Consequently, the tron parameters are equal to the arithmetic average sums of the parameters of their constituent toryces. In the reality-polarized toryces, the imaginary parameters are assumed to be real.

- Self-polarized trons are made up of self-polarized real toryces coexisting **concurrently** inside their trons, so the tron parameters are equal to the arithmetic sums of the parameters of their respective constituent toryces.

- The exponential excitation quantum states m of toryces forming trons depend on the spacetime levels L of the Multiverse to which they belong as shown in Table 7.4.

Table 7.4. Spacetime levels of the Multiverse.

Spacetime levels	Excitation quantum states m			
	Electron	**Positron**	**Singulatron**	**Ethertron**
L0	$m = 0$	$m = 0$	$m = 0$	$m = 0$
L1	$m = 1$	$m = 1$	$m = 0$	$m = 0$
L2	**$m = 2$**	**$m = 2$**	**$m = 1$**	**$m = 1$**
L3	$m = 3$	$m = 3$	$m = 2$	$m = 2$
L4	$m = 4$	$m = 4$	$m = 3$	$m = 3$
Lx	$m = x$	$m = x$	$m = x - 1$	$m = x - 1$
....	....	....	....	....

L2 = spacetime level of the ordinary matter.

Notes

Notes

<u>*Notes*</u>

8. Examples of Elementary Matter Particles

8.1 Reality-polarized harmonic lambda electron $e_{0,0,0}^{-1}$ is made up of two harmonic real negative toryces $E_{0,0,0}^{-\frac{1}{2}}$ and one half of the harmonic imaginary negative toryx $\breve{E}_{0,0,0}^{-2}$ as shown by the equation:

$$e_{0,0,0}^{-1} = 2E_{0,0,0}^{-\frac{1}{2}} + \frac{1}{2}\breve{E}_{0,0,0}^{-2} \tag{8.1-1}$$

Figure 8.1 and Table 8.1 show cross-section, dimensions, compositions and properties, of the reality-polarized harmonic lambda electron $e_{0,0,0}^{-1}$.

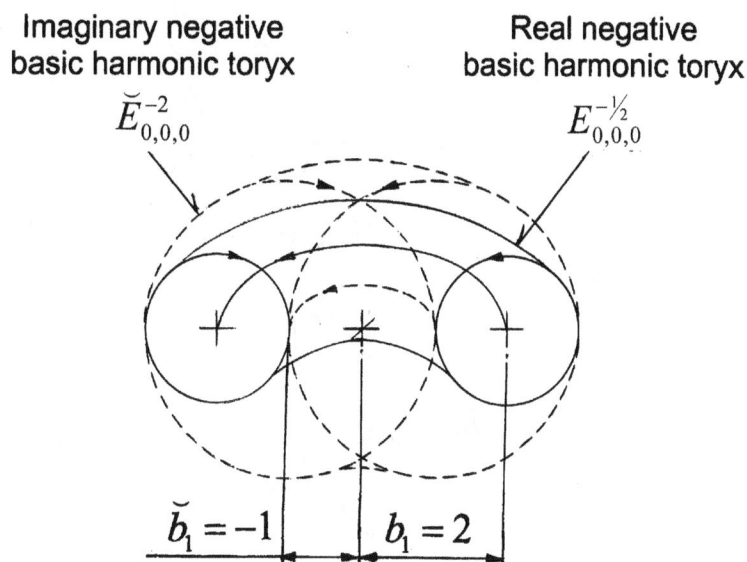

Figure 8.1. Cross-section and dimensions of the reality-polarized harmonic lambda electron $e_{0,0,0}^{-1}$ and its constituent toryces.

Table 8.1. Composition and properties of the reality-polarized harmonic lambda electron $e_{0,0,0}^{-1}$ and its constituent toryces.

Toryx	b_1	$\beta_1 = \beta_{2t}$	w_2	e_t/e	μ_t/μ_N	m_{tg}/m_e	P	N
$E_{0,0,0}^{-\frac{1}{2}}$	2.0	0.866025	1.154701	-0.50	- 5.80196032	0.500	-1.5	2
$\breve{E}_{0,0,0}^{-2}$	-1.0	-1.732051i	0.577350i	-2.00	- 23.20784127i	2.000	+6.0	0.5
Harmonic lambda electron $e_{0,0,0}^{-1}$				**-1.00**	**-11.60392064**	**1.000**	**0.00**	**1**

8.2 *Reality-polarized harmonic lambda positron* $e_{0,0,0}^{+1}$ is made up of two harmonic real positive toryces $E_{0,0,0}^{+\frac{1}{2}}$ and one half of the harmonic imaginary positive toryx $\breve{E}_{0,0,0}^{+2}$ as shown by the equation:

$$e_{0,0,0}^{+1} = 2E_{0,0,0}^{+\frac{1}{2}} + \tfrac{1}{2}\,\breve{E}_{0,0,0}^{+2} \qquad\qquad (8.2\text{-}1)$$

Figure 8.2 and Table 8.2 show cross-section, dimensions, compositions and properties of the reality-polarized harmonic lambda positron $e_{0,0,0}^{+1}$.

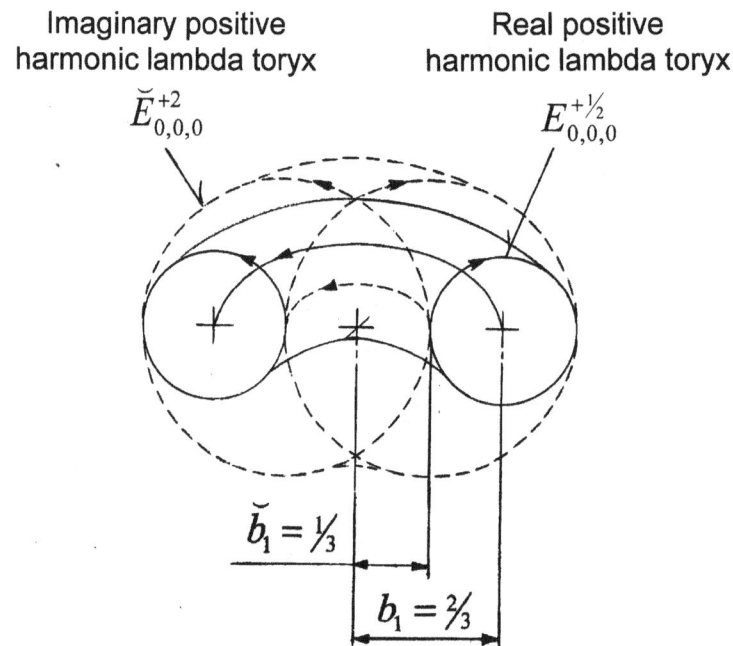

Figure 8.2. Cross-section and dimensions of the reality-polarized basic harmonic positron $e_{0,0,0}^{+1}$ and its constituent toryces.

Table 8.2. Composition and properties of the reality-polarized harmonic lambda positron $e_{0,0,0}^{+1}$ and its constituent toryces.

Toryx	b_1	$\beta_1 = \beta_{2t}$	w_2	e_t/e	μ_t/μ_N	m_{tg}/m_e	P	N
$E_{0,0,0}^{+\frac{1}{2}}$	$\tfrac{2}{3}$	0.866025	1.154701	+ 0.50	+1.93398677	0.500	$+\tfrac{1}{6}$	2
$\breve{E}_{0,0,0}^{+2}$	$\tfrac{1}{3}$	- 1.732051i	0.577350i	+ 2.00	+8.63594709i	2.000	$-\tfrac{2}{3}$	0.5
Harmonic lambda positron $e_{0,0,0}^{+1}$				**+1.00**	**+3.86797355**	**1.000**	**0.00**	**1**

8.3 *Vorticity-polarized harmonic lambda ethertron* $a_{0,0,0}^{0}$ is made up of one harmonic real negative toryx $A_{0,0,0}^{-1/2}$ and one harmonic real positive toryx $A_{0,0,0}^{+1/2}$ as shown by the equation:

$$a_{0,0,0}^{0} = A_{0,0,0}^{-1/2} + A_{0,0,0}^{+1/2} \qquad (8.3\text{-}1)$$

Figure 8.3 and Table 8.3 show cross-section, dimensions, compositions and properties of the harmonic lambda ethertron $a_{0,0,0}^{0}$.

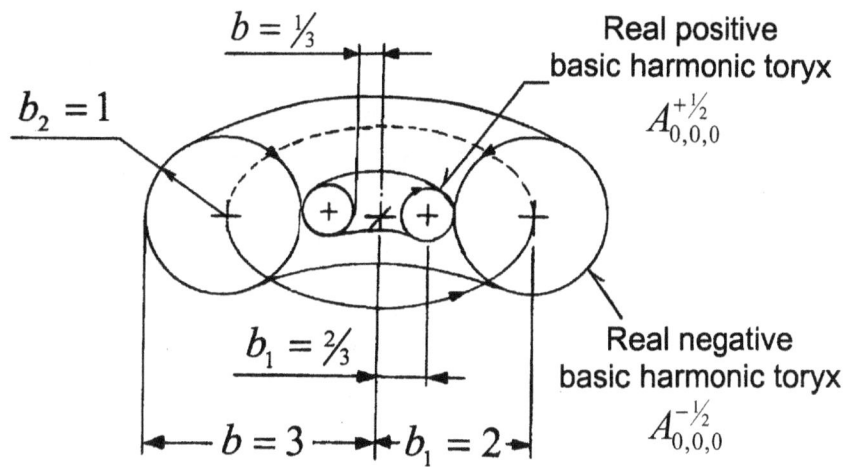

Figure 8.3. Cross-section and dimensions of the vorticity-polarized harmonic lambda ethertron $a_{0,0,0}^{0}$ and its constituent toryces.

Table 8.3. Composition and properties of the vorticity-polarized harmonic ethertron $a_{0,0,0}^{0}$ and its constituent toryces.

Toryx	b_1	$\beta_1 = \beta_{2t}$	w_2	e_t/e	μ_t/μ_N	m_{tg}/m_e	P	N
$A_{0,0,0}^{-1/2}$	2.0	0.866025	1.154701	-0.50	-5.80196032	0.500	-1.5	1
$A_{0,0,0}^{+1/2}$	$2/3$	0.866025	1.154701	+0.50	+1.93398677	0.500	$+1/6$	1
Harmonic lambda ethertron $a_{0,0,0}^{0}$				**0.00**	**-1.93398677**	**0.500**	$-2/3$	1

8.4 *Vorticity-polarized harmonic lambda singulatron* $\breve{a}^0_{0,0,0}$ is made up one harmonic imaginary negative toryx $\breve{A}^{-2}_{0,0,0}$ and one harmonic imaginary positive toryx $\breve{A}^{+2}_{0,0,0}$ as shown by the equation:

$$\breve{a}^0_{0,0,0} = \breve{A}^{-2}_{0,0,0} + \breve{A}^{+2}_{0,0,0} \tag{8.4-1}$$

Figure 8.4 and Table 8.4 show cross-section, dimensions, compositions and properties of the harmonic lambda singulatron $\breve{a}^0_{0,0,0}$.

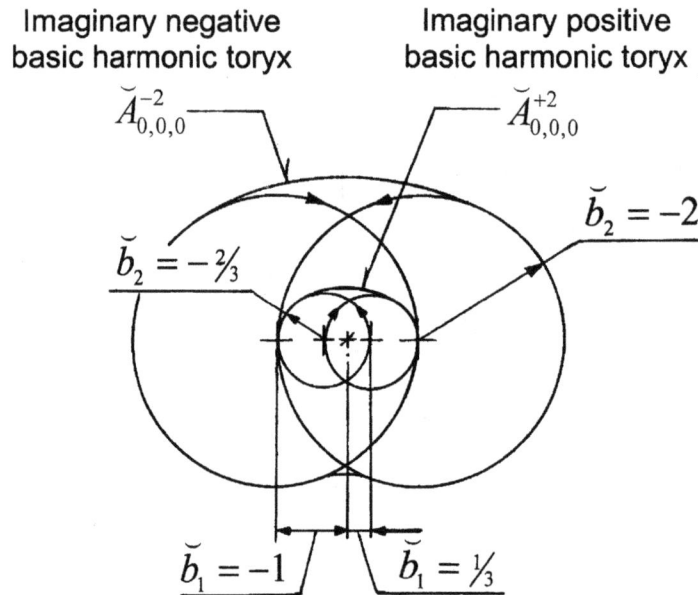

Figure 8.4 Cross-section and dimensions of the vorticity-polarized harmonic lambda singulatron $\breve{a}^0_{0,0,0}$ and its constituent toryces.

Table 8.4. Composition and properties of the vorticity-polarized harmonic lambda singulatron $\breve{a}^0_{0,0,0}$ and its constituent toryces.

Toryx	b_1	$\beta_1 = \beta_{2t}$	w_2	e_t/e	μ_t/μ_N	m_{tg}/m_e	P	N
$\breve{A}^{-2}_{0,0,0}$	-1.0	- 1.732051i	- 0.577350i	-2.0	-23.20784127i	2.000	+6.0	1
$\breve{A}^{+2}_{0,0,0}$	$\frac{1}{3}$	1.732051i	0.577350i	+2.0	+7.73594709i	2.000	$-\frac{2}{3}$	1
Harmonic lambda singulatron $\breve{a}^0_{0,0,0}$				**0.00**	**-7.73594709**	**2.000**	$+\frac{8}{3}$	**1**

8.5 Self-polarized unexcited harmonic electron $ae_{H,0,0}^{-1}$ is made up of one self-polarized real negative harmonic toryx $A_{H,0,0}^{-1/2}$ and one self-polarized real negative harmonic toryx $E_{H,0,0}^{-1/2}$ as given by the equation:

$$ae_{H,0,0}^{-1} = A_{H,0,0}^{-1/2} + E_{H,0,0}^{-1/2} \qquad (8.5\text{-}1)$$

Figure 8.5 and Table 8.5 show cross-section, dimensions, compositions and properties of the unexcited harmonic electron $ae_{H,0,0}^{-1}$.

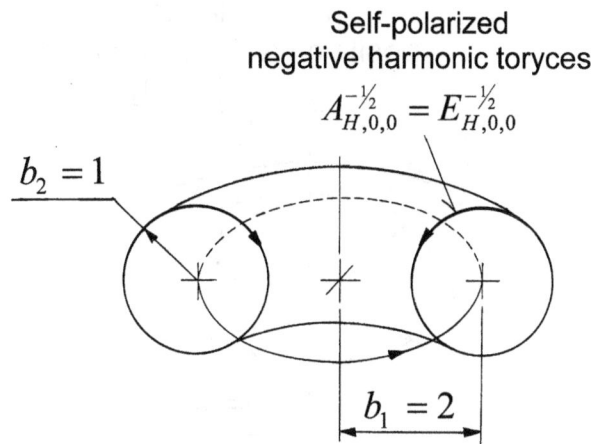

Figure 8.5. Cross-section and dimensions of the self-polarized unexcited harmonic electron $ae_{H,0,0}^{-1}$ and its constituent toryces.

Table 8.5. Composition and properties of the self-polarized unexcited harmonic electron $ae_{H,0,0}^{-1}$ and its constituent toryces.

Toryx	b_1	$\beta_1 = \beta_{2t}$	w_2	e_t/e	μ_t/μ_N	m_{tg}/m_e	N
$A_{H,0,0}^{-1/2}$	2.0	0.866025	1.154701	-0.50	-5.80196032	0.500	1
$E_{H,0,0}^{-1/2}$	2.0	0.866025	1.154701	-0.50	-5.80196032	0.500	1
Self-polarized harmonic electron $ae_{H,0,0}^{-1}$				**-1.00**	**-11.60392064**	**1.000**	**1**

8.6 *Self-polarized unexcited harmonic positron* $ae_{H,0,0}^{+1}$ is made up of one self-polarized real positive harmonic toryx $A_{H,0,0}^{+\frac{1}{2}}$ and one self-polarized positive harmonic toryx $E_{H,0,0}^{+\frac{1}{2}}$ as the equation:

$$ae_{H,0,0}^{+1} = A_{H,0,0}^{+\frac{1}{2}} + E_{H,0,0}^{+\frac{1}{2}} \qquad (8.6\text{-}1)$$

Figure 8.6 and Table 8.6 show cross-section, dimensions, compositions and properties of the unexcited harmonic positron $ae_{H,0,0}^{+1}$.

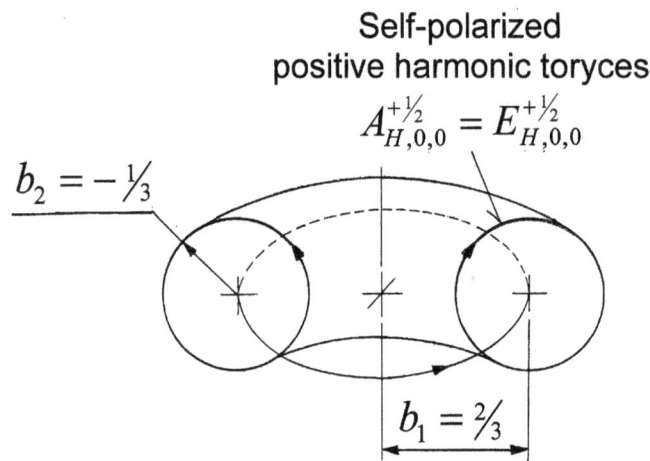

Figure 8.6. Cross-section and dimensions of the self-polarized unexcited harmonic positron $ae_{H,0,0}^{+1}$ and its constituent toryces.

Table 8.6. Composition and properties of the self-polarized unexcited harmonic positron $ae_{H,0,0}^{+1}$ and its constituent toryces.

Toryx	b_1	$\beta_1 = \beta_{2t}$	w_2	e_t/e	μ_t/μ_N	m_{tg}/m_e	N
$A_{H,0,0}^{+\frac{1}{2}}$	$\frac{2}{3}$	0.866025	1.154701	+0.50	+1.93398677	0.500	1
$E_{H,0,0}^{+\frac{1}{2}}$	$\frac{2}{3}$	0.866025	1.154701	+0.50	+1.93398677	0.500	1
Self-polarized harmonic positron $ae_{H,0,0}^{+1}$				**+1.00**	**+ 3.86797355**	**1.000**	**1**

8.7 *Self-polarized excited harmonic electron* $ae_{H,1,0}^{-1}$ is made up of one self-polarized real negative excited harmonic toryx $A_{H,1,0}^{-1/3}$ and one self-polarized real negative excited harmonic toryx $E_{H,1,0}^{-2/3}$ as given by the equation:

$$ae_{H,1,0}^{-1} = A_{H,1,0}^{-1/3} + E_{H,1,0}^{-2/3} \tag{8.7-1}$$

Figure 8.7 and Table 8.7 show cross-section, dimensions, compositions and properties of the unexcited harmonic electron $ae_{H,1,0}^{-1}$.

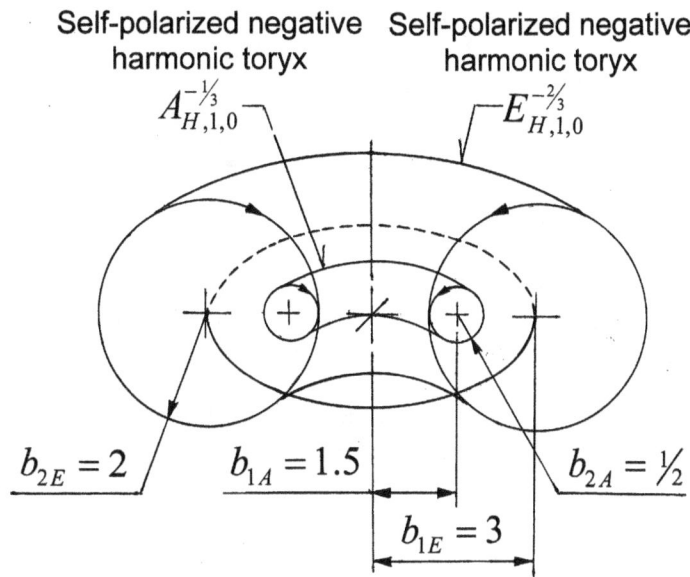

Figure 8.7. Cross-section and dimensions of the self-polarized excited harmonic electron $ae_{H,1,0}^{-1}$ and its constituent toryces.

Table 8.7. Composition and properties of the self-polarized real negative excited harmonic electron $ae_{H,1,0}^{-1}$ and its constituent toryces.

Toryx	b_1	$\beta_1 = \beta_{2t}$	w_2	e_t / e	μ_t / μ_N	m_{tg} / m_e	N
$A_{H,1,0}^{-1/3}$	1.5	0.942809	$1.060660i$	$-1/3$	-3.15818717	$1/3$	1
$E_{H,1,0}^{-2/3}$	3.0	0.745356	1.341641	$-2/3$	-9.98706475	$2/3$	1
Self-polarized excited electron $ae_{H,1,0}^{-1}$				**-1.00**	**-13.14525192**	**1.000**	**1**

8.8 Self-polarized excited harmonic positron $ae_{H,1,0}^{+1}$ is made up of one real self-polarized positive excited harmonic toryx $A_{H,1,0}^{+\frac{1}{3}}$ and one self-polarized real positive excited harmonic toryx $E_{H,1,0}^{+\frac{2}{3}}$ as given by the equation:

$$ae_{H,1,0}^{+1} = A_{H,1,0}^{+\frac{1}{3}} + E_{H,1,0}^{+\frac{2}{3}} \qquad (8.8\text{-}1)$$

Figure 8.8 and Table 8.8 show cross-section, dimensions, compositions and properties of the unexcited harmonic positron $ae_{H,1,0}^{+1}$.

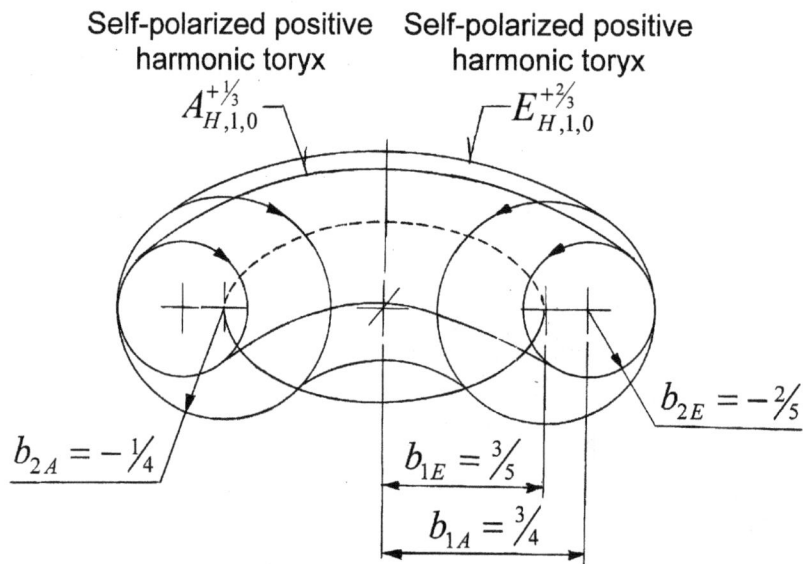

Figure 8.8. Cross-section and dimensions of the self-polarized excited harmonic positron $ae_{H,1,0}^{+1}$ and its constituent toryces.

Table 8.8. Composition and properties of the self-polarized excited harmonic positron $ae_{H,1,0}^{+1}$ and its constituent toryces.

Toryx	b_1	$\beta_1 = \beta_{2t}$	w_2	e_t/e	μ_t/μ_N	m_{tg}/m_e	N
$A_{H,1,0}^{+\frac{1}{3}}$	$\frac{3}{4}$	0.942809	1.060660	$+\frac{1}{3}$	+1.57909359	$\frac{1}{3}$	1
$E_{H,1,0}^{+\frac{2}{3}}$	$\frac{3}{5}$	0.745356	1.341641	$+\frac{2}{3}$	+1.99741295	$\frac{2}{3}$	1
Self-polarized excited positron $ae_{H,1,0}^{+1}$				**+1.00**	**+3.57650654**	**1.000**	**1**

8.9 Excited lambda electron $e_{m,n,q}^{-1}$ is made up of one excited real negative lambda toryx $E_{m,n,q}^{-}$ and one excited imaginary negative lambda toryx $\breve{E}_{m,n,q}^{-}$ as shown by the equation:

$$e_{m,n,q}^{-1} = E_{m,n,q}^{-} + \breve{E}_{m,n,q}^{-} \tag{8.9-1}$$

Figure 8.9 and Table 8.9 show cross-section, dimensions, compositions and properties of the excited lambda electron $e_{2,1,0}^{-1}$ corresponding to the spacetime level of atoms *L2* (ordinary matter).

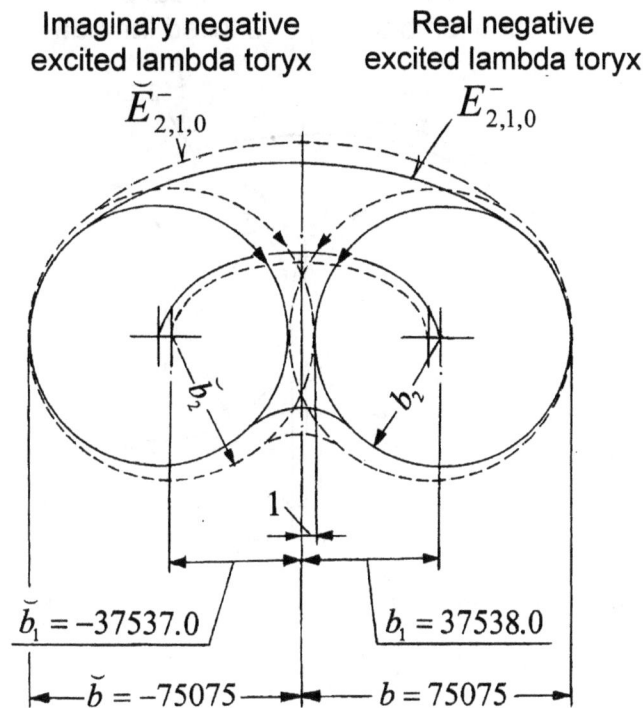

Imaginary negative excited lambda toryx $\breve{E}_{2,1,0}^{-}$ Real negative excited lambda toryx $E_{2,1,0}^{-}$

$\breve{b}_2$ b_2

1

$\breve{b}_1 = -37537.0$ $b_1 = 37538.0$

$\breve{b} = -75075$ $b = 75075$

Figure 8.9. Cross-section and dimensions of the excited lambda electron $e_{2,1,0}^{-1}$ of ordinary matter.

Table 8.9. Composition and properties of the excited lambda electron $e_{2,1,0}^{-1}$ of ordinary matter.

Toryx	b_1	e_t/e	μ_t/μ_N	m_{tg}/m_e	P	N
$E_{2,1,0}^{-}$	37538.0	-0.99997336	-1835.609190	0.99997336	-75075.0	1.00002664
$\breve{E}_{2,1,0}^{-}$	-37538.0	-1.00002664	-1835.706994i	1.00002664	+75075.0	0.99997336
Electron $e_{2,1,0}^{-1}$	- 1.0000000	-1835.658091	1.00000000	0.000	1

8.10 *Excited lambda positron* $e_{m,n,q}^{+1}$ is made up of one excited real positive lambda toryx $E_{m,n,q}^{+}$ and one excited imaginary positive lambda toryx $\breve{E}_{m,n,q}^{+}$ as shown by the equation:

$$e_{m,n,q}^{+1} = E_{m,n,q}^{+} + \breve{E}_{m,n,q}^{+} \tag{8.10-1}$$

Figure 8.10 and Table 8.10 show cross-section, dimensions, compositions and properties of the excited lambda positron $e_{2,1,0}^{+1}$ of ordinary matter.

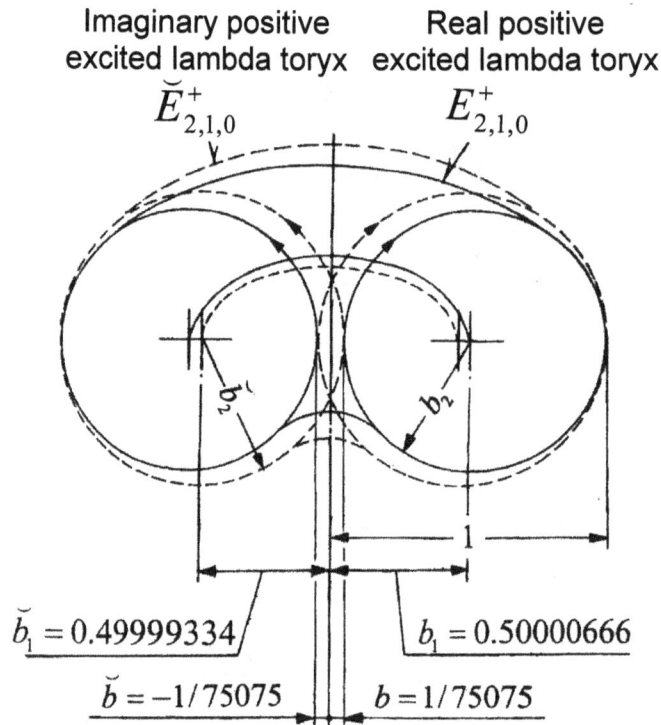

Figure 8.10. Cross-section and dimensions of the excited lambda positron $e_{2,1,0}^{+1}$ of ordinary matter.

Table 8.10. Composition and properties of the excited lambda positron $e_{2,1,0}^{+1}$ of ordinary matter.

Toryx	b_1	e_t/e	μ_t/μ_N	m_{tg}/m_e	P	N
$E_{2,1,0}^{+}$	0.50000666	+0.99997336	+0.024450	0.99997336	$+1.3320\times10^{-5}$	1.00002664
$\breve{E}_{2,1,0}^{+}$	0.49999334	+1.00002664	+0.024452i	1.00002664	-1.3320×10^{-5}	0.99997336
Positron $e_{2,1,0}^{+1}$		**+1.0000000**	**+0.024451**	**1.00000000**	**0.000**	**1**

8.11 *Excited lambda ethertron* $a^0_{m,n,q}$ is made up of one excited real negative lambda toryx $A^-_{m,n,q}$ and one excited real positive lambda toryx $A^+_{m,n,q}$ as shown by the equation:

$$a^0_{m,n,q} = A^-_{m,n,q} + A^+_{m,n,q} \qquad (8.11\text{-}1)$$

Figure 8.11 and Table 8.11 show cross-section, dimensions, compositions and properties of the excited lambda ethertron $a^0_{1,1,0}$ of ordinary matter.

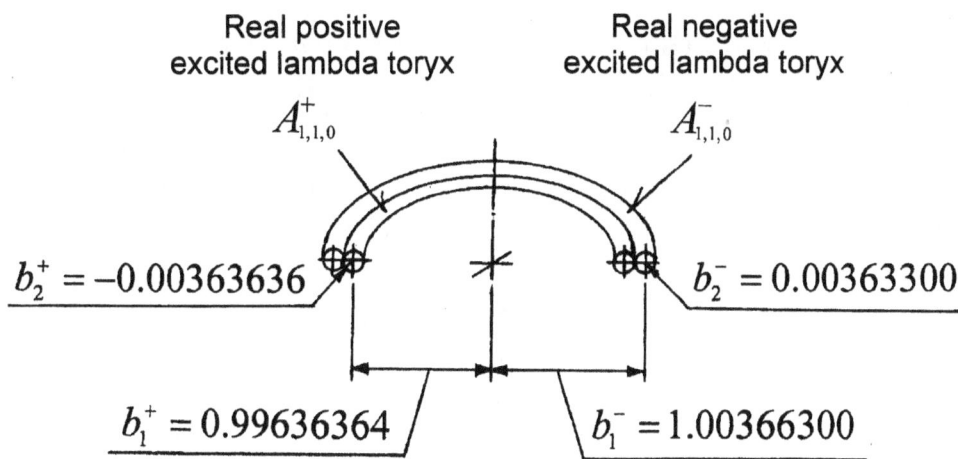

Real positive
excited lambda toryx

Real negative
excited lambda toryx

$A^+_{1,1,0}$

$A^-_{1,1,0}$

$b^+_2 = -0.00363636$

$b^-_2 = 0.00363300$

$b^+_1 = 0.99636364$

$b^-_1 = 1.00366300$

Figure 8.11. Cross-section and dimensions of the excited lambda ethertron $a^0_{1,1,0}$ of ordinary matter.

Table 8.11. Composition and properties of the excited lambda ethertron $a^0_{1,1,0}$ of ordinary matter.

Toryx	b_1	e_t/e	μ_t/μ_N	m_{tg}/m_e	P	N
$A^-_{1,1,0}$	1.00366300	- 0.00364964	-0.02454023	0.00364964	-0.00367637	1
$A^+_{1,1,0}$	0.99636364	+ 0.00364964	+0.02436175	0.00364964	+0.00362309	1
Ethertron $a^0_{1,1,0}$	**0.00**	**-0.00008924**	**0.00364964**	**-0.00002664**	**1**

8.12 *Excited lambda singulatron* $\breve{a}^{0}_{m,n,q}$ is made up of one excited imaginary negative lambda toryx $\breve{A}^{-}_{m,n,q}$ and one excited imaginary positive lambda toryx $\breve{A}^{+}_{m,n,q}$ as shown by the equation:

$$\breve{a}^{0}_{m,n,q} = \breve{A}^{-}_{m,n,q} + \breve{A}^{+}_{m,n,q} \tag{8.12-1}$$

Figure 8.12 and Table 8.12 show cross-section, dimensions, compositions and properties of the excited lambda singulatron $\breve{a}^{0}_{1,1,0}$ of ordinary matter.

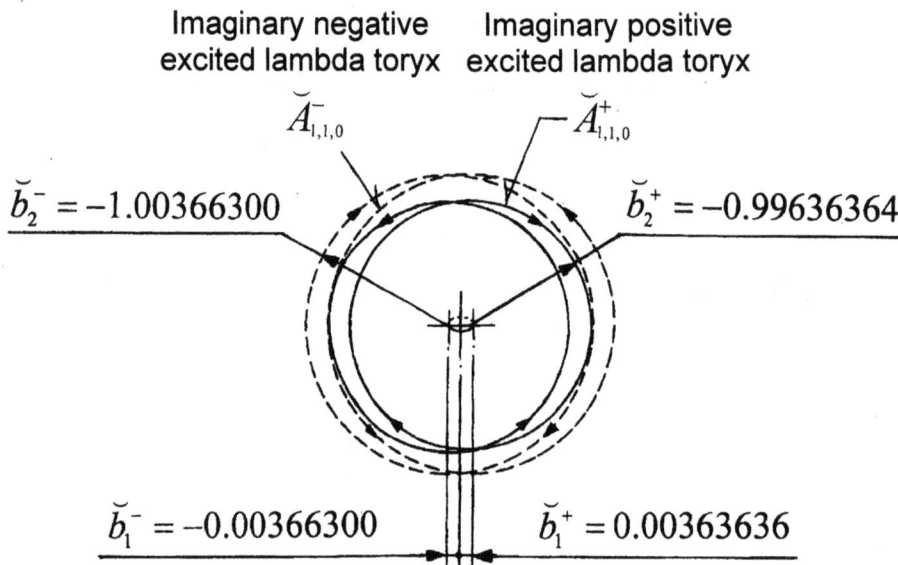

Figure 8.12. Cross-section and dimensions of the excited lambda singulatron $\breve{a}^{0}_{1,1,0}$ of ordinary matter.

Table 8.12. Properties of the excited lambda singulatron $\breve{a}^{0}_{1,1,0}$ of ordinary matter.

Toryx	b_1	e_t/e	μ_t/μ_N	m_{tg}/m_e	P	N
$\breve{A}^{-}_{1,1,0}$	0.00366300	- 273.998175	-1842.382113i	274.00	+276.007326	1
$\breve{A}^{+}_{1,1,0}$	0.00363636	+ 273.998175	+1828.882971i	274.00	-272.007273	1
Singulatron $\breve{a}^{0}_{1,1,0}$	**0.00**	**- 6.69957132**		**274.00**	**2.000027**	**1**

8.13 Combined harmonic lambda ce-tron $ce^0_{0,0,0}$ is made up of one harmonic lambda electron $e^{-1}_{0,0,0}$ and one harmonic lambda positron $e^{+1}_{0,0,0}$ as shown by the equation:

$$ce^0_{0,0,0} = e^{-1}_{0,0,0} + e^{+1}_{0,0,0} \tag{8.13-1}$$

Figure 8.13 and Table 8.13 show cross-section, dimensions, compositions and properties of the combined harmonic lambda ce-tron $ce^0_{0,0,0}$.

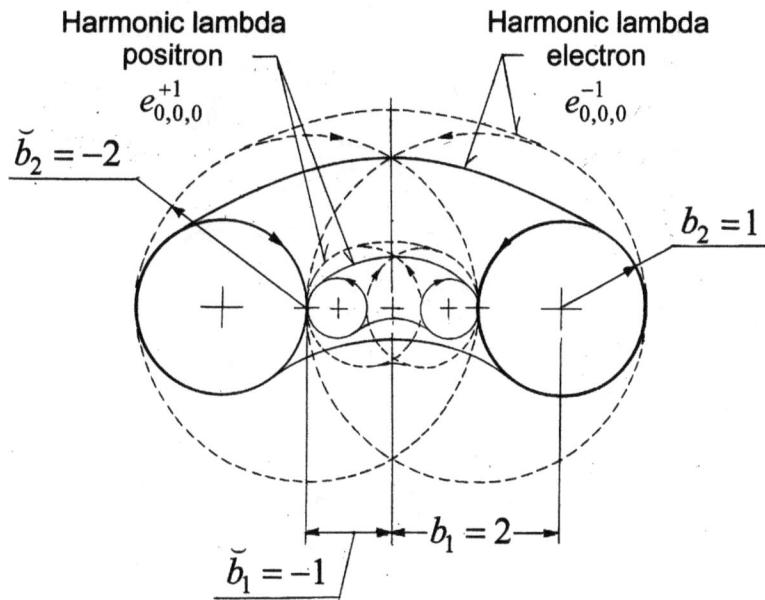

Figure 8.13. Cross-section and dimensions of properties of the combined harmonic lambda ce-tron $ce^0_{0,0,0}$ and its constituent toryces.

Table 8.13. Composition and properties of properties of the combined harmonic lambda ce-tron $ce^0_{0,0,0}$ and its constituent toryces.

Harmonic lambda trons		e_t/e	μ_t/μ_N	m_{tg}/m_e	P	N
Electron $e^{-1}_{0,0,0}$	Table 8.1	-1.00	-11.60392064	1.000	0.00	1
Positron $e^{+1}_{0,0,0}$	Table 8.2	+1.00	+3.86797355	1.000	0.00	1
Combined harmonic ce-tron $ce^0_{0,0,0}$		**0.00**	**-7.73594709**	**2.000**	**0.00**	**1**

8.14 *Combined harmonic lambda ca-tron* $ca^0_{0,0,0}$ is made up of one harmonic lambda singula-tron $\breve{a}^0_{0,0,0}$ and four harmonic lambda ethertrons $a^0_{0,0,0}$ as shown by the equation:

$$ca^0_{0,0,0} = \breve{a}^0_{0,0,0} + 4a^0_{0,0,0} \tag{8.14-1}$$

Figure 8.14 and Table 8.14 show cross-section, dimensions, compositions and properties of the combined harmonic lambda ca-tron $ca^0_{0,0,0}$.

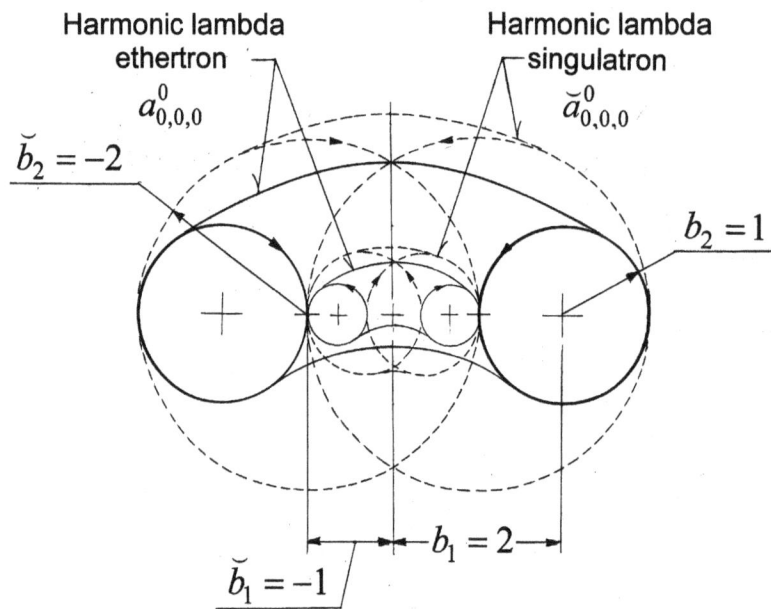

Figure 8.14. Cross-section and dimensions of properties of the combined harmonic lambda ca-tron $ca^0_{0,0,0}$ and its constituent toryces.

Table 8.14. Composition and properties of properties of the combined harmonic lambda ca-tron $ca^0_{0,0,0}$ and its constituent toryces.

Harmonic lambda trons		e_t/e	μ_t/μ_N	m_{tg}/m_e	P	N
Singulatron $\breve{a}^0_{0,0,0}$	Table 8.4	0.00	-7.73594709	2.000	$+\frac{8}{3}$	1
Ethertron $a^0_{0,0,0}$	Table 8.3	0.00	-1.93398677	0.500	$-\frac{2}{3}$	4
Combined harmonic ca-tron $ca^0_{0,0,0}$		**0.00**	**-7.73594709**	**2.000**	**0.00**	**1**

8.15 Combined excited lambda ce-tron $ce_{m,n,q}^{0}$ is made up one electron $e_{m,n,q}^{-1}$ and one positron $e_{m,n,q}^{+1}$ as shown by the equation and in Table 8.15:

$$ce_{m,n,q}^{0} = e_{m,n,q}^{-1} + e_{m,n,q}^{+1} \tag{8.15-1}$$

Table 8.15. Composition and properties of the combined excited lambda ce-tron $e_{2,1,0}^{-1}$ of ordinary matter.

Excited lambda trons		e_t/e	μ_t/μ_N	m_{tg}/m_e	P	N
Electron $e_{2,1,0}^{-1}$	Table 8.9	- 1.000000	-1835.658091	1.000000	0.0	1
Positron $e_{2,1,0}^{+1}$	Table 8.10	+1.000000	+0.024451	1.000000	0.0	1
Combined excited ce-tron $ce_{2,1,0}^{0}$	**0.000000**	**-1835.633640**	**2.000000**	**0.0**	**1**

8.16 Combined excited lambda ca-tron $ca_{m,n,q}^{0}$ is made up one singulatron $\breve{a}_{m,n,q}^{0}$ and N ethertrons $a_{m,n,q}^{0}$ as shown by the equation and in Table 8.16:

$$ca_{m,n,q}^{0} = \breve{a}_{m,n,q}^{0} + Na_{m,n,q}^{0} \tag{8.16-1}$$

Table 8.16. Composition and properties of the combined excited lambda ca-tron $ca_{m,n,q}^{0}$ of ordinary matter.

Excited lambda trons		e_t/e	μ_t/μ_N	m_{tg}/m_e	P	N
Singulatron $\breve{a}_{1,1,0}^{0}$	Table 8.12	0.00	- 6.69957132	274.000000	+2.000027	1
Ethertron $a_{1,1,0}^{0}$	Table 8.11	0.00	-0.00008924	0.00364964	-0.000027	75076
Combined excited ca-tron $ca_{1,1,0}^{0}$	**0.000000**	**- 6.69957132**	**274.000000**	**0.000000**	**1**

For a case in which the excitation quantum states n of the toryces forming a singulatron and N ethertrons correspond to the spacetime limits, the combined excited ca-tron $ca_{1,1,0}^{0}$ becomes the extreme form of ***ether*** (Fig. 8.16) in which one singulatron is surrounded by N ethertrons with N defined by the equation:

$$N = \left(\frac{1 \pm b_s}{b_s}\right)^2 \approx 1.511267 \times 10^{51} \tag{8.16-2}$$

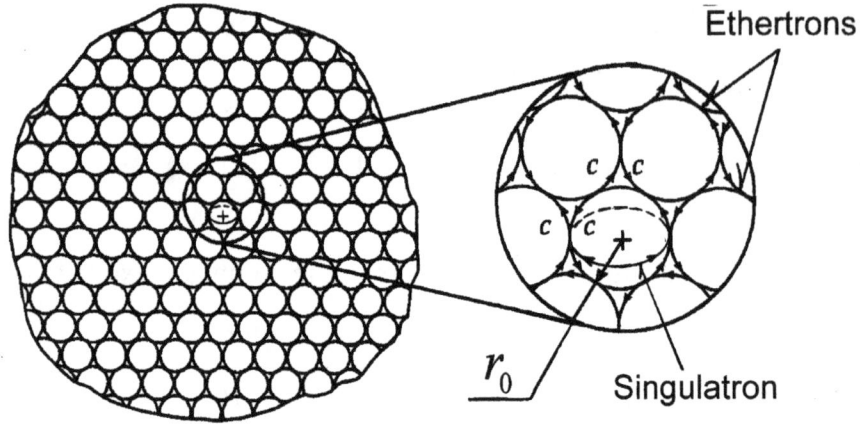

Figure 8.16. Extreme case of ether.

8.17 *Reality-polarized oscillated resonant harmonic lambda electron* $er_{0,0,1}^{-1}$ is made up of two real negative oscillated harmonic toryx $E_{0,0,1}^{-1/2}$ and two-third of the imaginary negative resonant oscillated harmonic toryx $Er_{0,0,1}^{-3/2}$ as shown by the equation:

$$er_{0,0,1}^{-1} = 2E_{0,0,1}^{-1/2} + \tfrac{2}{3}\, \breve{E}r_{0,0,1}^{-3/2} \tag{8.17-1}$$

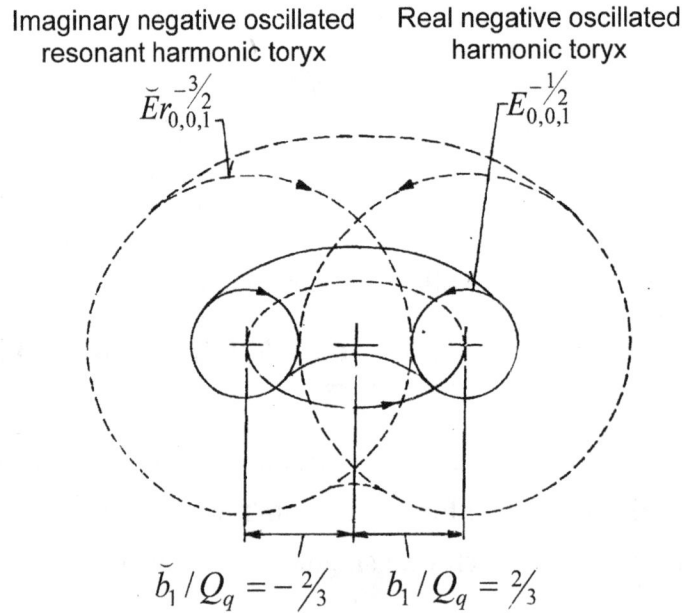

Figure 8.17. Cross-section and dimensions of the reality-polarized oscillated resonant harmonic lambda electron $er_{0,0,1}^{-1}$ and its constituent toryces.

Figure 8.17 and Table 8.17 show cross-section, dimensions, compositions and properties of the reality-polarized oscillated resonant harmonic lambda electron $er_{0,0,1}^{-1}$ and its constituent tory-ces.

Table 8.17. Composition and properties of the reality-polarized oscillated resonant harmonic lambda electron $er_{0,0,1}^{-1}$ and its constituent toryces.

Trons	b_1	$\beta_1 = \beta_{2t}$	w_2	e_t/e	μ_t/μ_N	m_{tg}/m_e	P	N
$E_{0,0,0}^{-1/2}$	2.0	0.866025	1.154701	-0.50	- 1.93398677	1.500	-4.5	1
$\breve{Er}_{0,0,1}^{-3/2}$	-2.0	-1.118034i	0.894427	-1.50	- 7.49029856	4.500	+22.5	1
Oscillated resonant harmonic electron $er_{0,0,1}^{-1}$				**-1.00**	**-4.71214267**	**3.000**	**9.00**	**1**

Notes

9. NUCLEONS, LIGHT ATOMS & ISOTOPES

Nucleons (protons and neutrons) are made up of three parts: ***nucleon crystal***, ***nucleon core*** and ***nucleon leptons.***

9.1 Nucleon Crystal

The main purpose of the nucleon crystal is to retain the excited and oscillated ethertrons and singulatrons, the constituents of the nucleon core. This is accomplished by locating the ethertrons and singulatrons inside the cavities (Fig. 9.1.1) formed around the center 1 and the vertices 2 through 9 of a nucleon bi-pyramid hexagonal crystal.

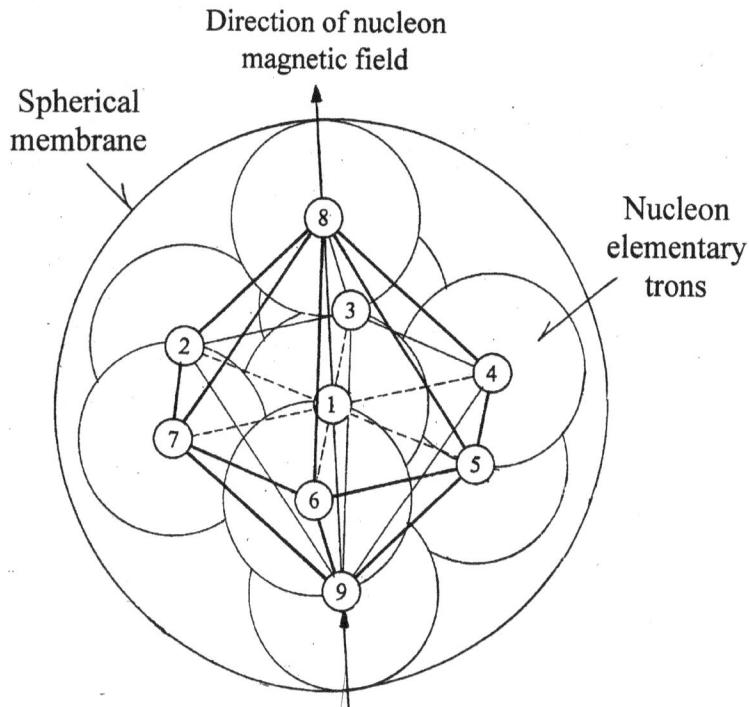

Figure 9.1.1. Isometric view of a nucleon crystal.

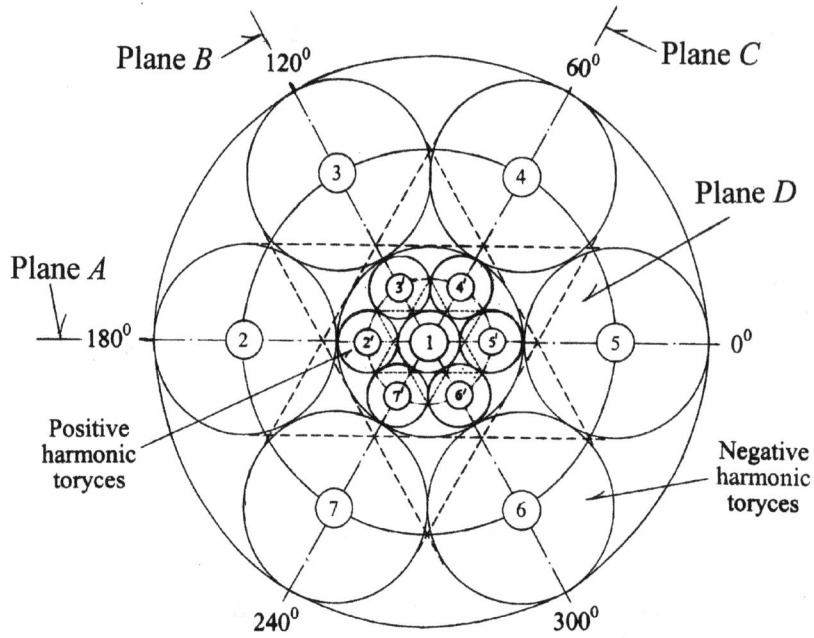

Figure 9.1.2. Cross-section of outer and inner parts of a nucleon crystal.

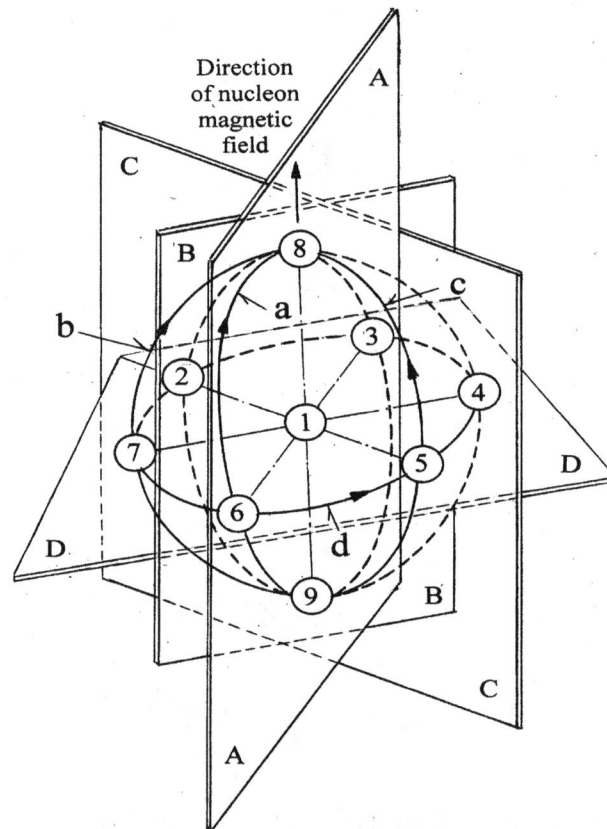

Figure 9.1.3. Isometric view of either outer or inner part of a nucleon crystal.

As shown in Fig. 9.1.2, 9.1.3 and Table 9.1, the nucleon crystal $\downarrow nx_{0,0,0}^{0}$ is made up of three combined harmonic ce-trons $ce_{0,0,0}^{0}$ and one combined harmonic ce-trons $\downarrow ce_{0,0,0}^{0}$. As shown in Table 8.13, each combined harmonic ce-tron $ce_{0,0,0}^{0}$ is made up of the harmonic lambda electron $e_{0,0,0}^{-1}$ and positron $e_{0,0,0}^{+1}$. Furthermore, as shown in Tables 8.1 and 8.2, each harmonic lambda electron $e_{0,0,0}^{-1}$ is made up of the real negative harmonic lambda toryx $E_{0,0,0}^{-\frac{1}{2}}$ and the imaginary negative harmonic lambda toryx $\breve{E}_{0,0,0}^{-2}$, while each harmonic lambda positron $e_{0,0,0}^{+1}$ is made up of the real positive harmonic lambda toryx $E_{0,0,0}^{+\frac{1}{2}}$ and the imaginary positive harmonic lambda toryx $\breve{E}_{0,0,0}^{+2}$.

Table 9.1. Composition and properties of the nucleon crystal $\downarrow nx_{0,0,0}^{0}$.

Combined harmonic trons	Tables	Planes in Figs. 9.1.2, 9.1.3	γ	$e_{,}/e$	μ/μ_N	m_g/m_e	P	N
Ce-tron $ce_{0,0,0}^{0}$	8.13	*A, B. C*	90^0	-1.00	0.00	2.00	0.0	3
Ce-tron $\downarrow ce_{0,0,0}^{0}$	8.13	*D*	90^0	+1.00	-7.73594709	2.00	0.0	1
Nucleon crystal $\downarrow nx_{0,0,0}^{0}$			-	**0.00**	**-7.73594709**	**8.00**	**0.0**	**1**

Three ce-trons $ce_{0,0,0}^{0}$ are located in the planes *A*, *B* and *C*, while one ce-tron $\downarrow ce_{0,0,0}^{0}$ is located in the plane *D*. The centers of all four ce-trons are located at the nucleon crystal center 1. The directions of the tron magnetic fields are perpendicular to the planes *A*, *B*, *C* and *D* in which they reside. A contribution of each ce-tron to the total magnetic moment of the nucleon crystal depends on the angle γ between the directions of their magnetic fields and the direction of the nucleon magnetic field shown in Figs. 9.1.1 and 9.1.3. Consequently, because for three ce-trons $ce_{0,0,0}^{0}$ the angle $\gamma = 90^0$, these ce-trons make no contribution to the magnetic moment of the nucleon crystal. At the same time, because for the ce-tron $\downarrow ce_{0,0,0}^{0}$ the angle $\gamma = 0^0$, this ce-tron makes a full contribution to the magnetic moment of the nucleon crystal.

Figure 9.1.2 shows a cross-section of the nucleon crystal $\downarrow nx_{0,0,0}^{0}$. It has two parts, outer and inner. The inner part is located inside the outer part and it is three times smaller than the outer part. Each part has the center 1 and eight vertices 2 through 9. As shown in Figs. 9.1.2 and 9.1.3, the vertices of the outer part are formed at the intersections of leading string strings *a*, *b*, *c* and *d* of the real negative harmonic toryces forming the harmonic electrons $e_{0,0,0}^{-1}$, while the vertices of the inner part are formed at the intersections of leading strings *a*, *b*, *c* and *d* of the real positive harmonic toryces forming the harmonic positrons $e_{0,0,0}^{+1}$.

9.2 Nucleon Core

As shown in Table 9.2, the nucleon core $\uparrow nc_{1,1,0}^{0}$ of the ordinary spacetime level *L2* is made up of three combined excited lambda ca-trons $\downarrow ca_{1,1,0}^{0}$ and four combined excited lambda ca-trons $\uparrow ca_{1,1,0}^{0}$. As shown in Table 8.16, each combined harmonic ca-tron $ca_{0,0,0}^{0}$ is made up of one imaginary excited lambda singulatron $\breve{a}_{1,1,0}^{0}$ and $N = 75076$ real excited lambda ethertrons $a_{1,1,0}^{0}$. Furthermore, as shown in Tables 8.11 and 8.12, each real excited lambda ethertron $a_{1,1,0}^{0}$ is made up of one real negative excited lambda toryx $A_{1,1,0}^{-}$ and one real positive excited lambda toryx $A_{1,1,0}^{+}$, while each imaginary excited lambda singulatron $\breve{a}_{1,1,0}^{0}$ is made up of one imaginary negative excited lambda toryx $\breve{A}_{1,1,0}^{-}$ and one imaginary positive excited lambda toryx $\breve{A}_{1,1,0}^{+}$.

Table 9.2. Composition and properties of the nucleon core $nc_{1,1,0}^{0}$ of the ordinary spacetime level *L2*.

Excited trons	Tables	Vertices in Figs. 9.1.1 – 9.1.3	Tables	μ / μ_N	m_g / m_e	P_t	N
Ca-tron $\downarrow ca_{1,1,0}^{0}$	8.16	1, 2, 5	8.16	-6.69957132	274.00	0.0	3
Ca-tron $\uparrow ca_{1,1,0}^{0}$	8.16	3, 4, 6, 7	8.16	+6.69957132	274.00	0.0	4
		Nucleon core $\uparrow nc_{1,1,0}^{0}$		**+6.69957132**	**1918.00**	**0.0**	**1**

9.3 Nucleons of the Ordinary Spacetime Level

Shown below are the structures and properties of proton and neutrons of the ordinary spacetime levels *L2*.

Proton $\uparrow p_{L2}^{+1}$ of the ordinary spacetime level *L2* is made up of one nucleon crystal $\downarrow nx_{0,0,0}^{0}$, one nucleon core $\uparrow nc_{1,1,0}^{0}$ and one harmonic lambda positron $\downarrow e_{0,0,0}^{+1}$ as shown in Table 9.3.1 and by the equation:

$$\uparrow p_{L2}^{+1} = \downarrow nx_{0,0,0}^{0} + \downarrow nc_{1,1,0}^{0} + \uparrow e_{0,0,0}^{+1} \tag{9.3-1}$$

Table 9.3.1. Components and properties of the proton $\uparrow p_{L2}^{+1}$ of the ordinary spacetime level *L2*.

Components	Tables	e_t/e	μ/μ_N	m_g/m_e	P_t	N
Excited nucleon crystal $\downarrow nx_{0,0,0}^0$	9.1	0.00	-7.73594709	8.00000	0.0	1
Nucleon core $\downarrow nc_{1,1,0}^0$	9.2	0.00	+6.69957132	1918.00000	0.0	1
Harmonic lambda positron $\uparrow e_{0,0,0}^{+1}$	8.2	+1.00	+3.86797355	1.000000	0.0	1
Proton $\uparrow p_{L2}^{+1}$		**+1.00**	**+2.83159778**	**1927.00000**	**0.0**	**1**
Measured values		+1.00	+ 2.79284736	1836.15267	-	-
Calculated/measured ratio		1.00	1.0139	1.0495	-	-

Unstable neutron $\downarrow nu_{L2}^0$ of the ordinary spacetime level *L2* is made up of one proton $\uparrow p_{L2}^{+1}$ and one resonant oscillated harmonic lambda electron $\downarrow er_{0,0,1}^{-1}$ as shown in Table 9.3.2 and by the equation:

$$\downarrow nu_{L2}^0 = \uparrow p_{l2}^{+1} + \downarrow er_{0,0,1}^{-1} \tag{9.3-2}$$

Table 9.3.2. Components and properties of the unstable neutron $\downarrow nu_{L2}^0$ of the ordinary spacetime level *L2*.

Components	Tables	e_t/e	μ/μ_N	m_g/m_e	P_t	N
Proton $\uparrow p_{L2}^{+1}$	9.3.1	+1.00	+2.83159778	1927.00000	0.0	1
Osc. res. harmonic electron $\downarrow er_{0,0,1}^{-1}$	8.17	-1.00	-4.71214267	3.0000000	9.0	1
Unstable neutron $\downarrow nu_{L2}^0$		**0.00**	**-1.88054489**	**1930.00000**	**9.0**	**1**
Measured values		0.00	-1.91304272	1838.68366	-	-
Calculated/measured ratio		1.00	0.98301249	1.0497	-	-

Stable neutron $\downarrow ns_{L2}^0$ of the ordinary spacetime level *L2* is made up of one proton $\uparrow p_{L2}^{+1}$ and one harmonic lambda electron $\downarrow e_{0,0,0}^{-1}$ as shown in Table 9.3.3 and by the equation:

$$\downarrow ns_{L2}^0 = \uparrow p_{L2}^{+1} + \downarrow e_{0,0,0}^{-1} \tag{9.3-3}$$

Table 9.3.3. Components and properties of the stable neutron $\downarrow n_{L2}^{0}$
of the ordinary spacetime level *L2*.

Components	Tables	e_t / e	μ / μ_N	m_g / m_e	P_t	N
Proton $\uparrow p_{L2}^{+1}$	9.3.1	+1.00	+2.83159778	1927.0000	0.0	1
Harmonic lambda electron $\downarrow e_{0,0,0}^{-1}$	8.1	-1.00	-11.60392064	1.0000	0.0	1
Stable neutron $\downarrow ns_{L2}^{0}$	**0.00**	**-8.77232286**	**1928.0000**	**0.0**	**1**

Figure 9.3 shows the nucleon crystal structure in which the small circles indicate locations of the center 1 and the vertices 2 through 7 occupied by the nucleon constituent trons, while the empty circles indicate locations of the vertices 8 and 9 occupied by constituent trons of adjacent nucleons in complex nuclei.

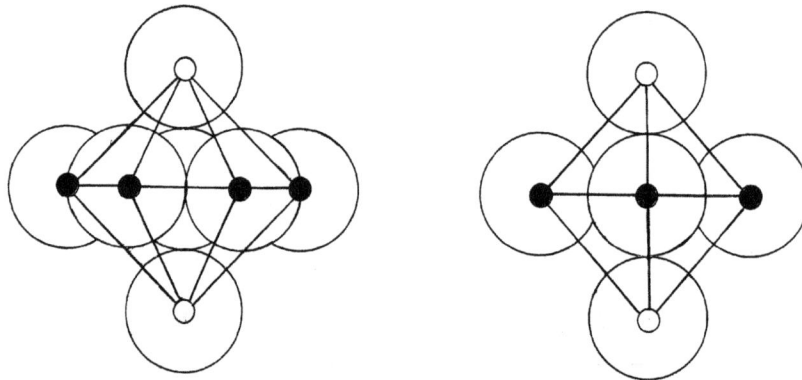

Figure 9.3. Front view (left) and side view (right) nucleon crystal structure.

9.4 Light Atoms & Isotopes of the Ordinary Spacetime Level

Shown below are the structures and properties of light atoms and isotopes of the ordinary spacetime levels *L2*.

Hydrogen atom $\downarrow _1^1 H_{L2}^{0}$ of the ordinary spacetime level *L2* is made up of one nucleon crystal $\downarrow nx_{0,0,0}^{0}$, one nucleon core $\uparrow nc_{1,1,0}^{0}$ and one combined excited ce-tron $\downarrow ce_{2,1,0}^{0}$ as shown in Table 9.4.1 and by the equation:

$$\downarrow _1^1 H_{L2}^{0} = \downarrow nx_{0,0,0}^{0} + \downarrow nc_{1,1,0}^{0} + \downarrow ce_{2,1,0}^{0} \qquad (9.4\text{-}1)$$

Table 9.4.1. Components and properties of hydrogen atom $\downarrow_1^1 H_{L2}^0$ of the ordinary spacetime level *L2*.

Components	Tables	e_t/e	μ/μ_N	m_g/m_e	P_t	N
Excited nucleon crystal $\downarrow nx_{0,0,0}^0$	9.1	0.00	-7.73594709	8.000000	0.0	1
Nucleon core $\downarrow nc_{1,1,0}^0$	9.2	0.00	+6.69957132	1918.000000	0.0	1
Combined excited ce-tron $\downarrow ce_{2,1,0}^0$	8.15	0.00	-1835.633640	2.000000	0.0	1
Hydrogen atom $\downarrow_1^1 H_{L2}^0$		**0.00**	**-1836.670016**	**1928.000000**	**0.0**	**1**
Measured values		0.00	-	1838.154061	-	-
Calculated/measured ratio		1.00	-	1.0489	-	-

Deuterium $\downarrow_1^2 H_{L2}^0$ of the ordinary spacetime level *L2* is made up of one hydrogen atom $\downarrow_1^1 H_{L2}^0$ and one stable neutron $\downarrow ns_{L2}^0$ as shown in Table 9.4.2 and by the equation:

$$\downarrow_1^2 H_{L2}^0 = \downarrow_1^1 H_{L2}^0 + \downarrow ns_{L2}^0 \tag{9.4-2}$$

Figure 9.4.1 shows a nucleon structure of the deuterium nucleus. Notably, when forming nuclei containing more than one nucleon, the adjacent nucleons become interlocked by filling vacant vertices of their nucleon crystals.

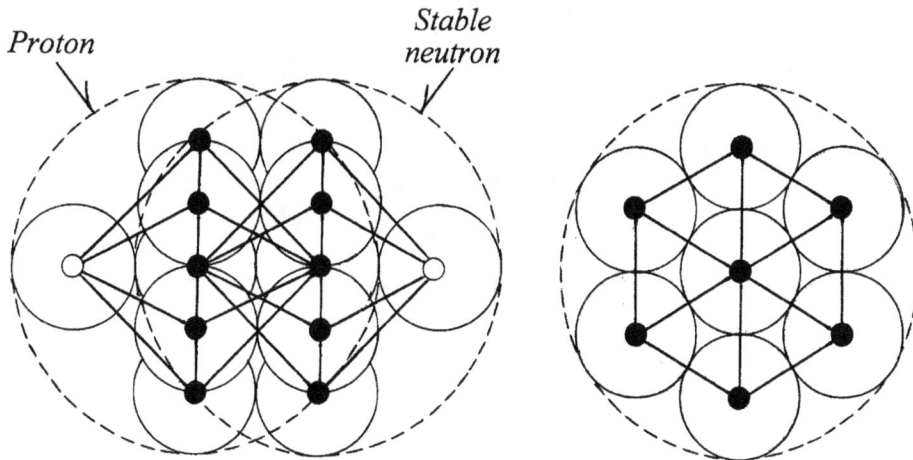

Figure 9.4.1. Structure of the deuterium nucleus:
front view (left) and side view (right).

Table 9.4.2. Components and properties of the deuterium $\downarrow_1^2 H_{L2}^0$

of the ordinary spacetime level *L2*.

Components	Tables	e_t/e	μ/μ_N	m_g/m_e	P_t	N
Hydrogen atom $\downarrow_1^1 H_{L2}^0$	9.4.1	0.00	-1836.670016	1928.000000	0.0	1
Stable neutron $\downarrow ns_{L2}^0$	9.3.3	0.00	-8.77232286	1928.000000	0.0	1
Deuterium $\downarrow_1^2 H_{L2}^0$		**0.00**	**-1845.442339**	**3856.000000**	**0.0**	**1**
Measured values		0.00	-	3671.485770	-	-
Calculated/measured ratio		1.00	-	1.0500	-	-

Tritium $\downarrow_1^3 H_{L2}^0$ of the ordinary spacetime level *L2* is made up of one hydrogen atom $\downarrow_1^1 H_{L2}^0$, one stable neutron $\downarrow ns_{L2}^0$ and one unstable neutron $\uparrow nu_{L2}^0$ as shown in Table 9.4.3, Fig. 9.4.2 and by the equation:

$$\downarrow_1^3 H_{L2}^0 = \downarrow_1^1 H_{L2}^0 + \downarrow ns_{L2}^0 + \uparrow nu_{L2}^0 \qquad (9.4\text{-}3)$$

Table 9.4.3. Components and properties of the tritium $\downarrow_1^3 H_{L2}^0$

of the ordinary spacetime level *L2*.

Components	Tables	e_t/e	μ/μ_N	m_g/m_e	P_t	N
Hydrogen atom $\downarrow_1^1 H_{L2}^0$	9.4.1	0.00	-1836.670016	1928.00000	0.0	1
Stable neutron $\downarrow ns_{L2}^0$	9.3.3	0.00	-8.77232286	1928.0000	0.0	1
Unstable neutron $\downarrow nu_{L2}^0$	9.3.2	0.00	-1.88084489	1930.0000	9.0	1
Tritium $\downarrow_1^3 H_{L2}^0$		**0.00**	**-1847.323184**	**5786.000000**	**9.0**	**1**
Measured values		-	-	5497.925226	-	-
Calculated/measured ratio		-	-	1.0524	-	-

Note: Tritium is unstable, because it contains the unstable neutron $\downarrow nu_{L2}^0$.

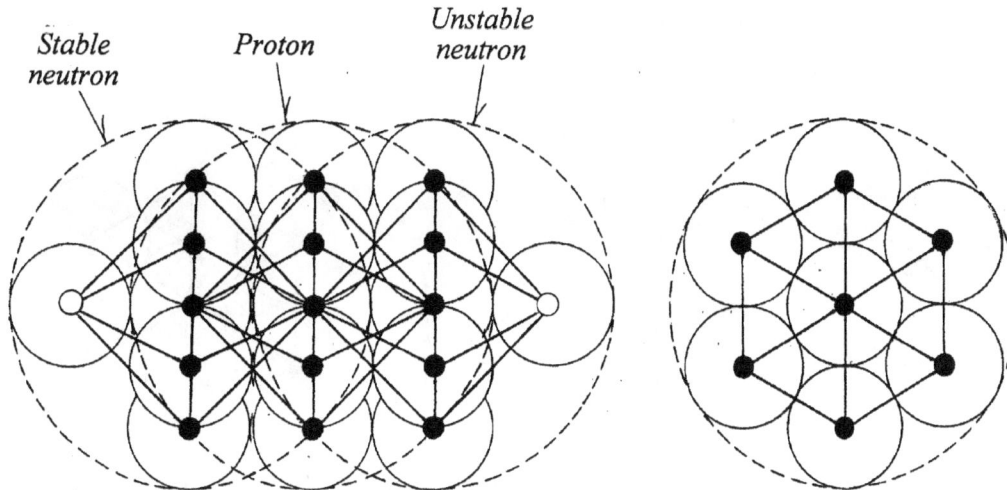

Figure 9.4.2. Structure of the tritium nucleus:
front view (left) and side view (right).

Helium-3 $\downarrow_2^3 He_{L2}^0$ of the ordinary spacetime level *L2* is made up of two hydrogen atoms $\downarrow_1^1 H_{L2}^0$ and $\uparrow_1^1 H_{L2}^0$ with opposite signs of their magnetic moments and one stable neutron $\downarrow ns_{L2}^0$ as shown in Table 9.4.4, Fig. 9.4.3 and by the equation:

$$\downarrow_2^3 He_{L2}^0 = \downarrow_1^1 H_{L2}^0 + \uparrow_1^1 H_{L2}^0 + \downarrow ns_{L2}^0 \qquad (9.4\text{-}4)$$

Table 9.4.4. Components and properties of the helium-3 $\downarrow_2^3 He_{L2}^0$ of the ordinary spacetime level *L2*.

Components	Table	e_t/e	μ/μ_N	m_g/m_e	P_t	N
Hydrogen atom $\downarrow_1^1 H_{L2}^0$	9.4.1	0.00	-1836.670016	1928.00000	0.0	1
Hydrogen atom $\uparrow_1^1 H_{L2}^0$	9.4.1	0.00	+1836.670016	1928.00000	0.0	1
Stable neutron $\downarrow ns_{L2}^0$	9.3.3	0.00	-8.77232286	1928.00000	0.0	1
Helium-3 $\downarrow_2^3 He_{L2}^0$		**0.00**	**- 8.77232286**	**5784.00000**	**0.0**	**1**
Measured values		-	-	5497.88877	-	-
Calculated/measured ratio		-	-	1.0520	-	-

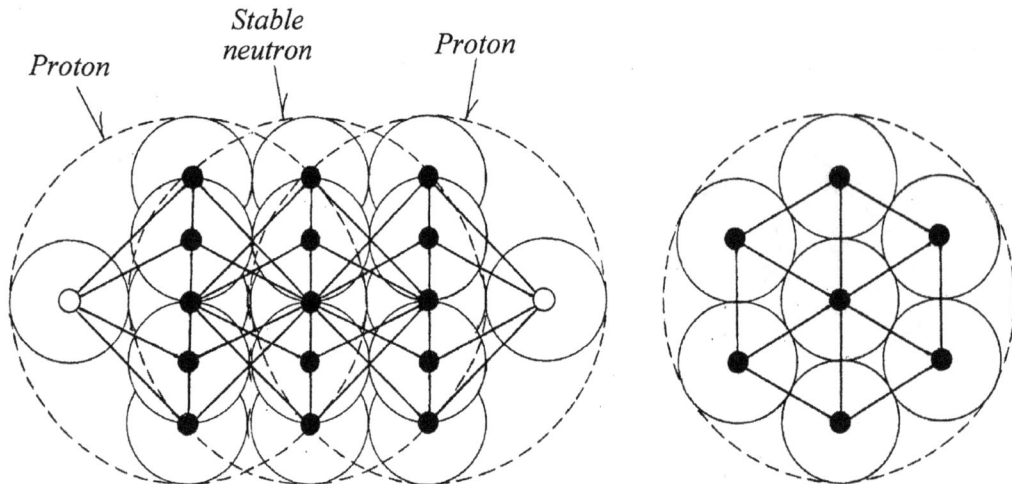

Figure 9.4.3. Structure of the helium-3 nucleus:
front view (left) and side view (right).

Helium-4 $_2^4He_{L2}^0$ of the spacetime level L is made up of two hydrogen atoms $_1^1H_{L2}^0$ and two stable neutrons ns_{L2}^0 with particles in both pairs having opposite signs of their magnetic moments as shown in Table 9.4.5, Fig. 9.4.4 and by the equation:

$$\downarrow_2^3He_{L2}^0 = \downarrow_1^1H_{L2}^0 + \uparrow_1^1H_{L2}^0 + \downarrow ns_{L2}^0 + \uparrow ns_{L2}^0 \qquad (9.4\text{-}5)$$

Table 9.4.5. Components and properties of the helium-4 $_2^4He_{L2}^0$
of the ordinary spacetime level $L2$.

Components	Table	e_t/e	μ/μ_N	m_g/m_e	P_t	N
Hydrogen atom $\downarrow_1^1H_{L2}^0$	9.4.1	0.00	-1836.670016	1928.00000	0.0	1
Hydrogen atom $\uparrow_1^1H_{L2}^0$	9.4.1	0.00	+1836.670016	1928.00000	0.0	1
Stable neutron $\downarrow ns_{L2}^0$	9.3.3	0.00	-8.77232286	1928.00000	0.0	1
Stable neutron $\uparrow ns_{L2}^0$	9.3.3	0.00	+8.77232286	1928.00000	0.0	1
Helium-4 $_2^4He_{L2}^0$		**0.00**	**0.00**	**7714.00000**	**0.0**	**1**
Measured values		-	-	7296.30308	-	-
Calculated/measured ratio		-	-	1.0572	-	-

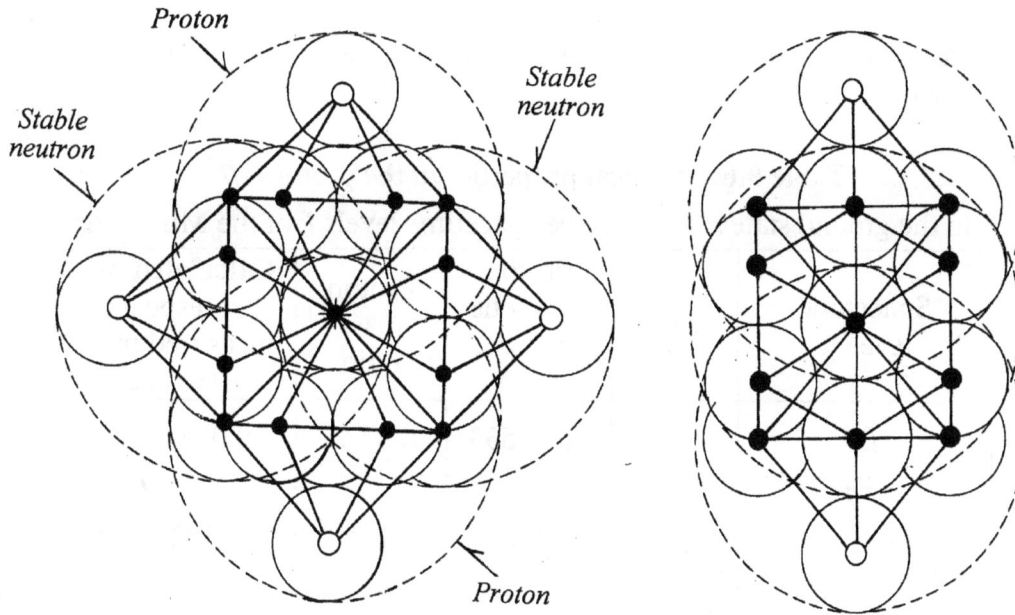

Figure 9.4.4 Structure of the helium-4 nucleus: front view (left) and side view (right).

9.5 Effect of Spacetime Levels of the Multiverse

Both the compositions and properties of hydrogen atoms and protons are greatly dependent on their spacetime levels L of the Multiverse.

Table 9.5 compares several physical properties of hydrogen atoms of the spacetime levels $L1$, $L2$ and $L3$ of the Multiverse.

Table 9.5. Relative parameters of the hydrogen atoms $\downarrow H_L^0$
in the ground state $n = 1$ of three spacetime levels L of the Multiverse.

Spacetime levels L of the Multiverse	Hydrogen atoms	Orbital electron radius ratio	Orbital electron magnetic moment ratio	Hydrogen atom mass ratio	Modulus of elasticity ratio
$L1$	$\downarrow H_{L1}^0$	1/137.25	1/11.37	1/37.08	3.53×10^8
$L2$ (ordinary space-time level)	$\downarrow H_{L2}^0$	**1.0**	**1.0**	**1.0**	**1.0**
$L3$	$\downarrow H_{L3}^0$	137.00	11.70	136.30	3.52×10^{-8}

As it follows from Table 9.5, as the spacetime level L of the Multiverse increases the orbital electron radius, the orbital electron magnetic moment and the mass of the hydrogen atom increase, while its modulus of elasticity decreases.

Table 9.6 compares several physical properties of protons of the spacetime levels *L1, L2* and *L3* of the Multiverse.

Table 9.6. Physical properties of the protons p_L^{+1}

in the ground state $n = 1$ of three spacetime levels *L* of the Multiverse.

Spacetime levels *L*	Proton	Proton magnetic moment ratio	Proton mass ratio	Calculated mass in respect to measured mass* GeV/c²
L1	↓ p_{L1}^{+1}	- 0.45533	1/37.78	0.024835152
L2 (ordinary space-time level)	↑ p_{L2}^{+1}	**1.0**	**1.0**	**0.938272046***
L3	↑ p_{L3}^{+1}	0.99986	136.365	127.9474747

As it follows from Table 9.6, as the spacetime level *L* of the Multiverse increases the proton magnetic moment increases and reaches its maximum value for the spacetime level *L2*. It is then decreases slightly as the spacetime level *L* increases. The proton mass increases with the increase of the spacetime level *L*.

The right column of Table 9.6 shows calculated proton masses of the spacetime levels *L1* and *L3* in respect to the measured proton mass* of the spacetime levels *L2* based on the proton mass ratios shown in the previous column. Notably, for the spacetime level *L3*, the calculated proton mass is equal to 127.9474747 GeV/c² that is 2.28% greater than the mass 125.09 ± 0.24 GeV/c² of the Higgs boson reported by CERN in 2017.

Notes

<u>*Notes*</u>

10. OSCILLATED ELEMENTARY PARTICLES

10.1 Leptons

According to the proposed theory, leptons belong to a group of particles containing oscillated electron and positron with masses significantly exceeding the mass of non-oscillated electron and positron. The leptons can be of two kinds: ***excited lambda leptons*** and ***harmonic leptons***. Both these leptons exist in the oscillation quantum states described in Section 6.4 .

Figure 10.1 shows a plot of a natural logarithm of the toryx oscillation factor $\ln Q_q$ as a function of the toryx oscillation quantum states q calculated from Eq. (6.4-1) in application to the excited lambda leptons of the ordinary matter $L2$.

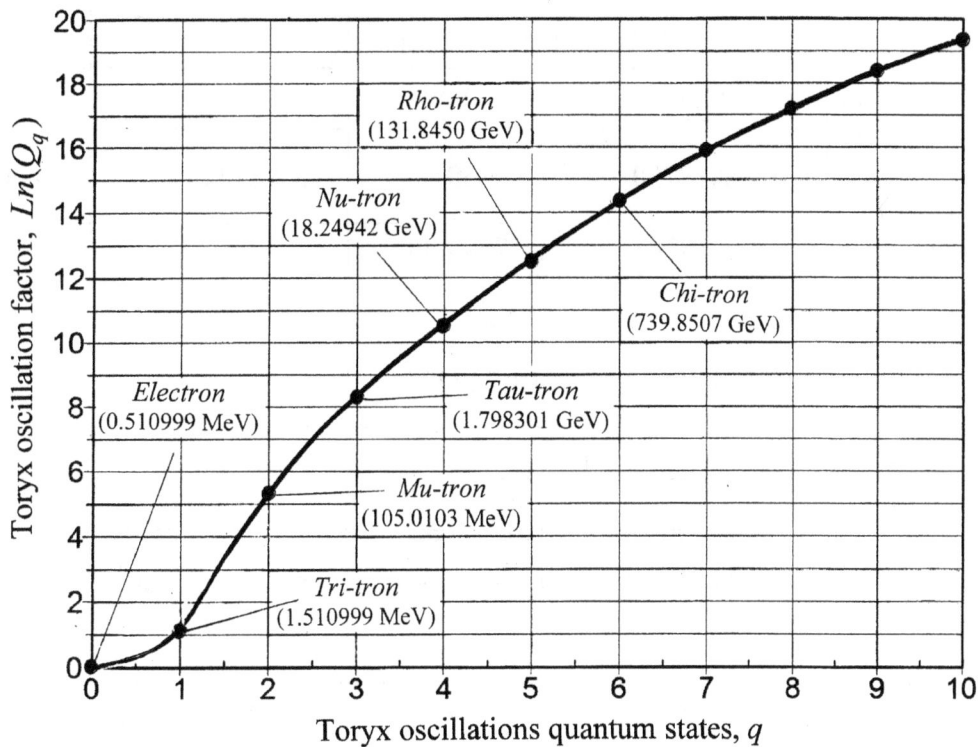

Figure 10.1. Toryx oscillation factor Q_q as a function of the toryx oscillation quantum states q.

Table 10.1 shows a comparison of calculated & measured properties of non-oscillated and oscillated excited lambda electrons and positrons (leptons) of the ordinary matter $L2$.

Table 10.1. Calculated & measured properties of non-oscillated and oscillated excited lambda electrons and positrons (leptons) of the ordinary matter *L2* ($b_1^+ = 0.50000666, \breve{b}_1^+ = 0.499999334, b_1^- = 37538.0, \breve{b}_1^- = -37537.0$).

Leptons		Positive and negative leptons		μ_t / μ_B for leptons	
Name	Symbol	$m_{tg}/m_e = Q_q$	m_{tg}, GeV/c^2	Positive	Negative
$e-tron$ $q=0$	$e_{2,1,0}^{\pm 1}$	**1.000000**	**0.00051099893**	**+1.331643 × 10⁻⁵**	**-0.99973052**
Measured values:		1.000000	0.00051099893	-	-1.00115965
Calc./measured ratio:		1.0000	1.00	-	0.9986
$3e-tron$ $q=1$	$e_{2,1,1}^{\pm 1}$	**3.000000**	**0.00151099893**	**+4.438808 × 10⁻⁶**	**-0.33324351**
Measured values:		-	-	-	-
Calc./measured ratio:		-	-	-	-
$\mu-tron$ $q=2$	$e_{2,1,2}^{\pm 1}$	**205.500000**	**0.1050102797**	**+6.480012 × 10⁻⁸**	**-4.864869 × 10⁻³**
Measured values		-206.768284	0.1056583668	-	-4.841970 × 10⁻³
Calc./measured ratio		0.9939	0.9939	-	1.0047
$\tau-tron$ $q=3$	$e_{2,1,3}^{\pm 1}$	**3519.18750**	**1.798301041**	**+3.783049 × 10⁻⁹**	**-2.840799 × 10⁻⁴**
Measured values		3478.18283	1.776990000	-	-
Calc./measured ratio		1.0120	1.0120	-	-
$\nu-tron$ $q=4$	$e_{2,1,4}^{\pm 1}$	**35713.23612**	**18.2494254**	**+3.728709 × 10⁻¹⁰**	**-2.799328 × 10⁻⁵**
Measured values		-	-	-	-
Calc./measured ratio		-	-	-	-
$\rho-tron$ $q=5$	$e_{2,1,5}^{\pm 1}$	**258014.1805**	**131.8449697**	**+5.161122 × 10⁻¹¹**	**-3.874712 × 10⁻⁶**
Measured values		244795.0427	125.0900000	-	-
Calc./measured ratio		1.0540 ?	1.0540 ?	-	-
$\chi-tron$ $q=6$	$e_{2,1,6}^{\pm 1}$	**1447851.734**	**739.8506841**	**+9.197368 × 10⁻¹²**	**-6.904923 × 10⁻⁷**
Measured values		1468.713450	750 ?	-	-
Calc./measured ratio		0.98646758 ?	0.98646758 ?	-	-

It is follows from Table 10.1 and Figure 10.1:

- Calculated masses of *electron, mu-tron,* and *tau-tron* are very close to their respective measured values.
- Calculated magnetic moments of *electron* and *mu-tron* are very close to their respective measured values.
- Calculated mass of *rho-tron* is 5.4% greater than the mass of Higgs boson measured at CERN in 2017.

10.2 Basic Quarks

Quarks are the elementary particles having fractional charges. We proposed that they are made up of harmonic oscillated toryces. We consider below two groups of quarks:

Group 1 includes the quarks with the fractional charge $+\frac{2}{3}e$. In the quarks of this group the relative radii of leading strings b_1 of one of their constituent toryces changes from 0.5 to $+\infty$.

Group 2 includes the quarks with the fractional charge $-\frac{1}{3}e$. In the quarks of this group the relative radii of leading strings b_1 of one of their constituent toryces changes from -0.5 to $-\infty$.

Table 10.2. provides a comparison of calculated and measured relative masses of quarks. The empty cells indicate masses of predicted still undiscovered quarks.

Table 10.2. Comparison of calculated and measured relative masses of quarks.

q	Quark name	Quark group 1: charge $+\frac{2}{3}e$		Quark name	Quark group 2: charge $-\frac{1}{3}e$	
		Ranges of relative masses m_g/m_e			Ranges of relative masses m_g/m_e	
		Calculated	Measured		Calculated	Measured
0	-	0.34 - 0.85	-	-	0.34 - 1,36	-
1	Up	1.02 - 2.56	1.80 – 3.00	Down	1.02 - 4.09	4.5 - 5.3
2	-	70.0 - 175.0	-	Strange	70.0 – 280.0	90 – 100.0
3	Charm	1199 - 2997	1250 -1300	Bottom	1199 - 4798	4215 - 4690
4	-	12166 - 30416	-	-	12166 - 48665	-
5	Top	87897 - 219742	171099 - 174430	-	87897 - 351586	-

10.3 Mesons

We proposed that mesons are made up of oscillated harmonic trons. Tables below show proposed compositions of several kinds of mesons and a comparison between their calculated and measured relative masses.

Light unflavored mesons - Table 10.3.1 shows compositions and masses of the light unflavored mesons of the ordinary matter *L2*.

Table 10.3.1. Compositions and properties of the light unflavored mesons of the ordinary matter *L2*.

Constituent harmonic trons of mesons and their quantities					Meson relative mass m_g/m_e		
π^0	$\downarrow a_{H,1,2}^0$	$\uparrow a_{H,1,2}^0$	-	-	Calc.	Measured	Ratio
	1	1	-	-	274.00	264.14263	1.0373
$\pi^\pm$	$\downarrow a_{H,1,2}^0$	$\uparrow a_{H,1,2}^0$	$\updownarrow ae_{0,1,1}^{\pm1}$	-	Calc.	Measured	Ratio
	1	1	1	-	277.00	273.13204	1.0142
η	$\downarrow a_{H,1,2}^0$	$\uparrow a_{H,1,2}^0$	-	-	Calc.	Measured	Ratio
	2	2	-	-	1096.000	1072.139	1.022
$\rho(770)$	$\downarrow a_{H,1,2}^0$	$\uparrow a_{H,1,2}^0$	-	-	Calc.	Measured	Ratio
	3	3	-	-	1644.000	1517.596	1.083

Strange mesons - Table 10.3.2 shows compositions and masses of the strange mesons of the ordinary matter *L2*.

Table 10.3.2. Compositions and properties of the strange mesons of the ordinary matter *L2*.

Constituent harmonic trons of mesons and their quantities					Meson relative mass m_g/m_e		
$K^\pm$	$\downarrow a_{H,1,2}^0$	$\uparrow a_{H,1,2}^0$	$\updownarrow ae_{H,1,2}^{\pm1}$	-	Calc.	Measured	Ratio
	3	3	1	-	1027.000	966.102	1.064
K^0	$\downarrow a_{H,1,2}^0$	$\uparrow a_{H,1,2}^0$	$\uparrow ae_{H,1,2}^{+1}$	$\downarrow ae_{H,1,1}^{-1}$	Calc.	Measured	Ratio
	3	3	1	1	1030.500	973.806	1.058

Charmed mesons - Table 10.3.3 shows compositions and masses of the charmed mesons of the ordinary matter *L2*.

Table 10.3.3. Compositions and properties of the charmed mesons of the ordinary matter *L2*.

	Constituent harmonic trons of mesons and their quantities				Meson relative mass m_g/m_e		
$D^{\pm}$	$\downarrow a_{H,1,2}^0$	$\updownarrow ae_{0,1,3}^{\pm1}$	$\downarrow a_{H,1,1}^0$	$\uparrow a_{H,1,1}^0$	Calc.	Measured	Ratio
	1	1	3	3	3668.188	3658.755	1.003
D^0	$\downarrow a_{H,1,2}^0$	$\uparrow ae_{H,1,3}^{+1}$	$\downarrow ae_{0,1,0}^{-1}$	-	Calc.	Measured	Ratio
	1	1	1	-	3657.188	3649.401	1.002

Charm strange, mesons - Table 10.3.4 shows compositions and masses of the charmed, strange mesons of the ordinary matter *L2*.

Table 10.3.4. Compositions and properties of the charmed, strange mesons of the ordinary matter *L2*.

	Constituent harmonic trons of mesons and their quantities				Meson relative mass m_g/m_e		
$D_s^{\pm}$	$\downarrow a_{H,0,2}^0$	$\uparrow a_{H,0,2}^0$	$\updownarrow ae_{H,1,3}^{\pm1}$	-	Calc.	Measured	Ratio
	2	2	1	-	3930.188	3852.239	1.020
$D_s^{*\pm}$	$\downarrow a_{H,0,2}^0$	$\uparrow a_{H,0,2}^0$	$\updownarrow ae_{H,1,3}^{\pm1}$	-	Calc.	Measured	Ratio
	3	3	1	-	4135.688	4133.668	1.001
$D_{s0}^{*\pm}$	$\downarrow a_{H,1,2}^0$	$\uparrow a_{H,1,2}^0$	$\updownarrow ae_{H,1,3}^{\pm1}$	-	Calc.	Measured	Ratio
	4	4	1	-	4615.188	4535.822	1.018
$D_{s1}^{\pm}$	$\downarrow a_{H,3,2}^0$	$\uparrow a_{H,3,2}^0$	$\updownarrow ae_{H,1,3}^{\pm1}$	-	Calc.	Measured	Ratio
	4	4	1	-	4834.388	4813.317	1.004
$D_{s2}^{\pm}$	$\downarrow a_{H,2,2}^0$	$\uparrow a_{H,2,2}^0$	$\updownarrow ae_{H,1,3}^{\pm1}$	-	Calc.	Measured	Ratio
	5	5	1	-	5060.438	5034.453	1.005

Cc̄ **mesons** - Table 10.3.5 shows compositions and masses of the $c\bar{c}$ mesons of the ordinary matter *L2*.

Table 10.3.5. Compositions and properties of the $c\bar{c}$ mesons
of the ordinary matter *L2*.

Constituent harmonic trons of mesons and their quantities					Meson relative mass m_g/m_e		
$\eta_c(1S)$	$\downarrow a_{H,1,3}^0$	$\uparrow a_{H,1,3}^0$	$\uparrow ae_{H,1,2}^{+1}$	$\downarrow ae_{H,1,2}^{-1}$	Calc.	Measured	Ratio
	1	1	3	3	5925.188	5832.302	1.016
$J/\psi(1S)$	$\downarrow a_{H,2,3}^0$	$\uparrow a_{H,2,3}^0$	$\uparrow ae_{H,1,2}^{+1}$	$\downarrow ae_{H,1,2}^{-1}$	Calc.	Measured	Ratio
	1	1	2	2	6100.781	6060.514	1.007
$\chi_{c0}(1S)$	$\downarrow a_{H,1,3}^0$	$\uparrow a_{H,1,3}^0$	$\uparrow ae_{H,1,2}^{+1}$	$\downarrow ae_{H,1,2}^{-1}$	Calc.	Measured	Ratio
	1	1	5	5	6747.250	6682.499	1.010
$\chi_{c1}(1S)$	$\downarrow a_{H,2,3}^0$	$\uparrow a_{H,2,3}^0$	$\uparrow ae_{H,1,2}^{+1}$	$\downarrow ae_{H,1,2}^{-1}$	Calc.	Measured	Ratio
	1	1	4	4	6922.781	6870.191	1.008

Bottom mesons - Table 10.3.6 shows compositions and masses of the bottom mesons of the ordinary matter *L2*.

Table 10.3.6. Compositions and properties of the bottom mesons
of the ordinary matter *L2*.

Constituent harmonic trons of mesons and their quantities				Meson relative mass m_g/m_e		
$B^{\pm}$	$\downarrow a_{H,0,3}^0$	$\uparrow a_{H,0,3}^0$	$\updownarrow ae_{H,1,3}^{\pm1}$	-	Calc.	Measured	Ratio
	2	2	1	-	10557.56	10331.04	1.022
B^0	$\downarrow a_{H,0,3}^0$	$\uparrow a_{H,0,3}^0$	$\uparrow ae_{H,1,3}^{+1}$	$\downarrow ae_{H,1,0}^{-1}$	Calc.	Measured	Ratio
	2	2	1	1	10558.56	10331.78	1.022

Bottom, charmed mesons - Table 10.3.7 shows compositions and masses of the bottom, charmed mesons of the ordinary matter *L2*.

Table 10.3.7. Compositions and properties of the bottom, charmed mesons of the ordinary matter *L2*.

	Constituent harmonic trons of mesons and their quantities				Meson relative mass m_g/m_e		
B_c^+	$\downarrow a_{H,0,3}^0$	$\uparrow ae_{H,1,3}^{+1}$	-	-	Calc.	Measured	Ratio
	5	1	-	-	12317.16	12281.83	1.003
B_c^-	$\downarrow a_{H,0,3}^0$	$\uparrow ae_{H,1,3}^{-1}$	-	-	Calc.	Measured	Ratio
	5	1	-	-	12317.16	12281.83	1.003

$B\bar{b}$ mesons - Table 10.3.8 shows compositions and masses of the $b\bar{b}$ mesons of the ordinary matter *L2*.

Table 10.3.8. Compositions and properties of the $b\bar{b}$ mesons of the ordinary matter *L2*.

	Constituent harmonic trons of mesons and their quantities				Meson relative mass m_g/m_e		
Y(1S)	$\downarrow a_{H,6,3}^0$	$\uparrow a_{H,6,3}^0$	$\uparrow ae_{H,1,2}^{+1}$	$\downarrow ae_{H,1,1}^{-1}$	Calc.	Measured	Ratio
	3	3	1	1	18684.2344	18513.3461	1.009
Y(2S)	$\downarrow a_{H,6,3}^0$	$\uparrow a_{H,6,3}^0$	$\uparrow ae_{H,1,2}^{+1}$	$\downarrow ae_{H,1,1}^{-1}$	Calc.	Measured	Ratio
	3	3	3	3	19708.7344	19615.0314	1.005
Y(3S)	$\downarrow a_{H,7,3}^0$	$\uparrow a_{H,7,3}^0$	$\uparrow ae_{H,1,2}^{+1}$	$\downarrow ae_{H,1,1}^{-1}$	Calc.	Measured	Ratio
	3	3	4	4	20412.0000	20264.6218	1.007

<u>*Notes*</u>

Notes

<u>*Notes*</u>

11. TORYCES OF THE MACRO-WORLD

In previous chapters we described a role of the *micro-toryces* in the formation of matter particles and atoms of the micro-world. In the macro-world, the assemblies of atoms contained in each body form the macro-spacetimes called the *macro-toryces* that become integral parts of each body.

11.1 Basic Structure & Parameters of Macro-Toryces

Figure 11.1 shows a macro-toryx associated with the central body A and encompassing the satellite body B.

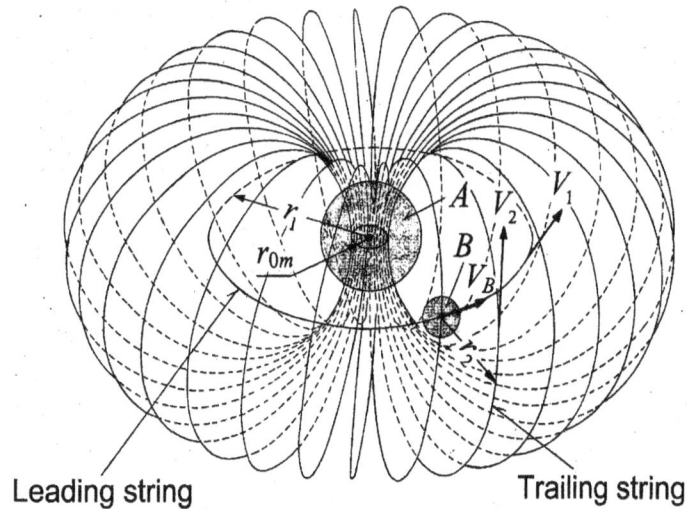

Figure 11.1. Central body A with a macro-toryx encompassing the satellite body B.

Similarly to the micro-toryx (Figs. 1.1.1 – 1.1.3), the macro-toryx contains two strings, a *leading string* and a *trailing string*. Leading string is double-circular; it is moving with the velocity V_1 along a circle with the radius r_1. Trailing string is double-toroidal with the radius r_2 in which each branch propagates along its toroidal spiral path with the spiral velocity V_2 that has two components, the translational velocity V_{2t} and the rotational velocity V_{2r}.

11.2 Macro-Toryx Spacetime Postulates

Exhibit 11.2.1 shows the spacetime postulates expressed in absolute spacetime units. These postulates are the same as those used for micro-toryces shown in Exhibit 1.3.

Exhibit 11.2.1. Macro-toryx spacetime postulates in absolute units.

- The length of one winding of trailing string L_2 is equal to the length of one winding of leading string L_1:

$$L_2 = L_1 = 2\pi r_1 \qquad (11.2\text{-}1)$$

- The macro-toryx eye radius r_{0m} is constant:

$$r_{0m} = r_1 - r_2 = const. \qquad (11.2\text{-}2)$$

- The spiral velocity of trailing string V_2 is constant at each point of its spiral path:

$$V_2 = \sqrt{V_{2t}^2 + V_{2r}^2} = c = const. \qquad (11.2\text{-}3)$$

Similarly to the micro-toryx, the spacetime postulates of the macro-toryces can also be simplified (Exhibit 11.2.2) by expressing the macro-toryx spacetime parameters in relative units in respect to the constant macro-toryx parameters: the eye radius r_{0m} and the velocity of light c.

Exhibit 11.2.2. Toryx spacetime postulates in relative units.

- The relative length of one winding of trailing string l_2 is equal to the relative length of one winding of leading string l_1:

$$l_2 = l_1 \qquad (11.2\text{-}4)$$

- The macro-toryx relative eye radius b_{0m} is equal to 1:

$$b_{0m} = b_1 - b_2 = 1 \qquad (11.2\text{-}5)$$

- The relative spiral velocity of trailing string β_2 is equal to 1 at each point of its spiral path:

$$\beta_2 = \sqrt{\beta_{2t}^2 + \beta_{2r}^2} = 1 \qquad (11.2\text{-}6)$$

11.3 Classification of Macro-Toryces & Macro-Trons

Similarly to the classification of the micro-toryces shown in Chapter 3, the macro-toryces are also divided into four groups as shown in Fig. 11.3:

- Real negative macro-toryces
- Real positive macro-toryces
- Imaginary positive macro-toryces
- Imaginary negative macro-toryces.

Similarly to the classification of the micro-trons shown in Section 7.3, the macro-trons are also divided into four groups:

- Macro-electrons
- Macro-positrons
- Macro-ethertrons
- Macro-singulatrons.

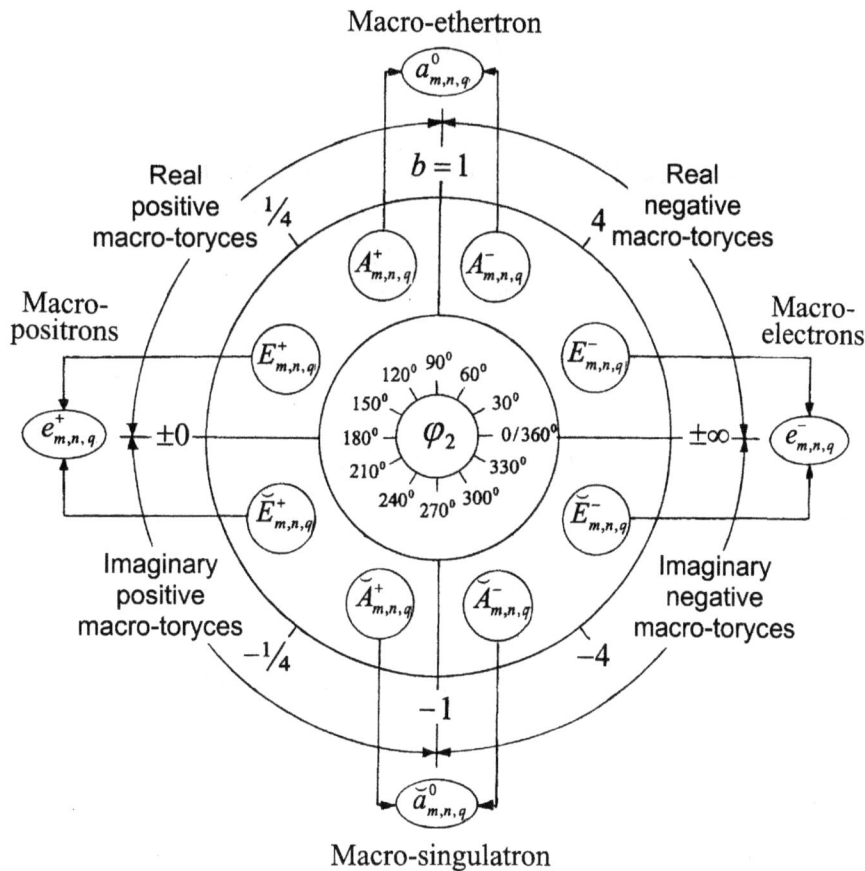

Figure 11.3. Formation of four macro-trons from polarized macro-toryces.

11.4 Macro-Toryx Law of Planetary Motion

We described In Section 1.7, the law of planetary system applicable to the micro-toryces by the equation:

$$\beta_1 = \frac{V_1}{c} = \frac{\sqrt{2b_1 - 1}}{b_1} \tag{11.4-1}$$

For the case when $b_1 \gg 1$, the above equation reduces to the form:

$$\beta_1 = \frac{V_1}{c} = \sqrt{\frac{2}{b_1}} \tag{11.4-2}$$

The above two equations also describe the **macro-toryx law of planetary motion** with only one difference. In the planetary system applicable to the micro-toryces, b_1 is expressed by Eq. (1.6-1a), whereas in the macro-toryx law of planetary motion b_1 is expressed by the equation:

$$b_1 = \frac{r_1}{r_{0m}} \tag{11.4-3}$$

where r_{0m} is the eye radius of the macro-toryx associated with the body A with the mass m_A (Fig. 11.1) that is equal to:

$$r_{om} = \frac{m_A G}{2c^2}, \tag{11.4-4}$$

where G is the Newtonian constant of gravitation.

From Eqs. (11.4-1), (11.4-3) and (11.4-4) and by considering that the velocity of the macro-toryx leading string V_1 is related to its period T_1 by the equation:

$$V_1 = 2\pi r_1 / T_1, \tag{11.4-5}$$

it is possible to express the macro-toryx law of planetary motion in the form:

$$r_1^3 = k T_1^2 \left(1 - \frac{r_{0m}}{2r_1} \right) \tag{11.4-6}$$

where the constant k is equal to:

$$k = \frac{r_{0m} c^2}{2\pi^2} = \frac{m_A G}{4\pi^2} \tag{11.4-7}$$

Consequently, for the case when r_1 is much greater than the macro-toryx eye radius r_{0m}, Eq. (11.4-6) reduces to the Kepler's third law of planetary motion:

$$r_1^3 = kT_1^2 \qquad (11.4\text{-}8)$$

11.5 Interaction Between Celestial Bodies

We proposed that the macro-trons associated with celestial bodies are responsible for the interactions between these bodies. Figure 11.5 shows two adjacent celestial bodies A and B and the respective macro-toryces A and B associated with these bodies.

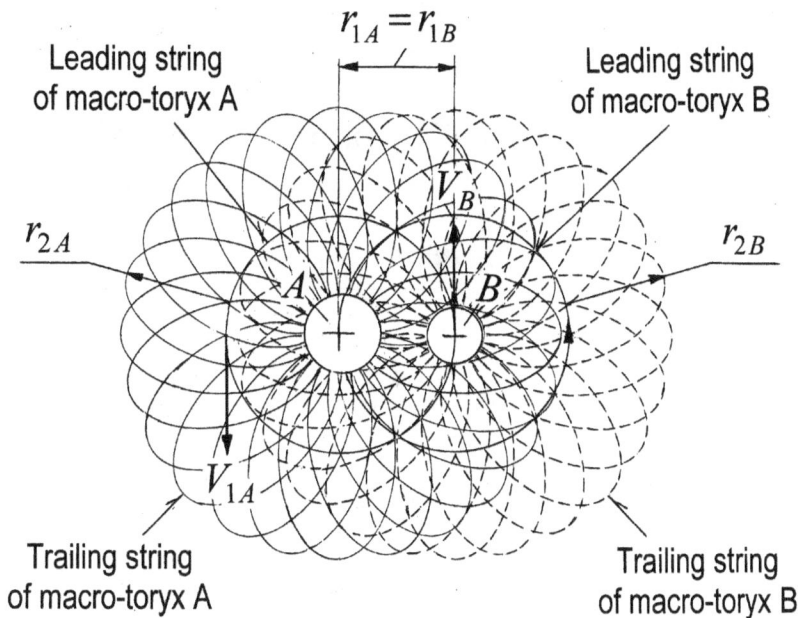

Figure 11.5. Two adjacent celestial bodies A and B with macro-toryces A and B associated with these bodies.

Let the macro-toryx A to be associated with the body A having the mass m_A. The macro-toryx circular leading string with the radius r_{1A} propagates around a center of the body A with the orbital velocity V_{1A}, while its toroidal trailing string with the radius r_{2B} propagates synchronously with the leading string. Located at the distance r_{1A} from the center of the body A is a center of the body B having the mass m_B. The body B moves around the center of the body A along a circular path with the radius $r_{1B} = r_{1A}$ with the orbital velocity V_B.

The behavior of the body B depends on a relationship between its orbital velocity V_B and the spiral velocity of leading string V_{1A} of the macro-toryx A associated with the body A. When these velocities are not exactly the same, it means the body B does not follow the macro-toryx law of planetary motion described by Eq. (11.4-1). Consequently, the body B will move in a radial direction either towards to or away from the body A with the *spacetime acceleration* a_{bs} described by the proposed equation shown in Exhibit 11.5.

Exhibit 11.5. Spacetime acceleration of a body.

Spacetime acceleration a_{bs} of the body B in respect to a center of the body A is equal to:

$$a_{bs} = \frac{V_{1A}^2 - V_B^2}{r_{1A}} = \frac{V_{1A}^2(1 - \gamma_V^2)}{r_{1A}} \qquad (11.5\text{-}1)$$

where γ_V is the velocity ratio that is equal to:

$$\gamma_V = V_B/V_1 \qquad (11.5\text{-}2)$$

Considering Eqs. (11.4-1), (11.4-4) and (11.5-1), the spacetime acceleration a_{bs} of the body B in respect to the body A separated by the distance b_{1A} is equal to:

$$a_{bs} = \frac{m_A G}{r_{1A}^2} \frac{2b_{1A} - 1}{2b_{1A}} (1 - \gamma_V^2) \qquad (11.5\text{-}3)$$

For a particular case when $b_{1A} \gg 1$ and $\gamma_V = 0$, Eq. (11.5-3) reduces to the form:

$$a_b = \frac{m_A G}{r_{1A}^2} \qquad (11.5\text{-}4)$$

Based on Eqs. (11.5-1) - (11.5-4), we may conclude:

- When the orbital velocity V_B of the body B around a center of the body A is the same as the velocity of propagation of leading string V_{1A} of the macro-toryx associated with the body A ($\gamma_V = 1$), the body B will orbit the body A according to the spacetime law of planetary motion and, consequently, the velocity V_B and its distance to the body B will remain unchanged.
- When the velocity V_B of the body B around the body A is different than the velocity of propagation of leading string V_1 of the macro-toryx associated with the body A, the body B accelerates either towards to or away from the body A.

11.6 Spacetime Law of Gravitation

Consider a case when the satellite body B with the mass m_B does not comply with the spacetime law of planetary motion and moves freely either towards to or away from the body A with the acceleration a_b. Let the dimensions of the body B to be negligibly small in comparison with its distance from the body A. In that case, the body B will not experience any force applied to it until after its free motion is affected by either internal or external causes. In that moment a holding force will be applied to the body B.

This holding force is called the ***spacetime gravitational force*** F_{gs}, and it is equal to the product of the body mass m_B and its acceleration a_{bs} as described by the equation:

$$F_{gs} = m_B a_{bs} \qquad (11.6\text{-}1)$$

Similarly to Eq. (7.1-5) applied to a toryx, when $Q_q = 1$, the inertial mass of the body B is equal to:

$$m_{Bi} = m_B \frac{2(b_{1A} - 1)}{2b_{1A} - 1} \qquad (11.6\text{-}2)$$

Consequently, we obtain from Eqs. (11.5-3), (11.6-1) and (11.6-2) that the spacetime gravitational force F_{gs} is equal to:

$$F_{gs} = \frac{m_A m_B G}{r_{1A}^2} \frac{(b_{1A} - 1)}{b_{1A}} (1 - \gamma_V^2) \qquad (11.6\text{-}3)$$

We called Eq. (11.6-3) the ***spacetime law of gravitation***. For a particular case when $b_1 \gg 1$ and $\gamma_V^2 \ll 1$, Eq. (11.6-3) reduces to a familiar equation expressing the Newton's universal law of gravitation:

$$F_{gN} = \frac{m_A m_B G}{r_{1A}^2} \qquad (11.6\text{-}4)$$

Based on Eqs. (11.5-1) - (11.5-3), (11.6-3) and Fig. 11.6, we may conclude:

- When the velocity V_B of the satellite body B around the central body A is the same as the velocity of propagation of the leading string V_{1A} of the macro-toryx associated with the central body A ($\gamma_V = 1$), the body B will orbit the body A according to the spacetime law of planetary motion and, consequently, the velocity V_B and its distance to the satellite body B will remain unchanged.
- When the velocity V_B of the satellite body B around the central body A is different than the velocity of propagation of the leading string V_{1A} of the macro-toryx associated with the central body A, the satellite body B accelerates either towards to or away from the body A.

- If the radius of the satellite body B is negligibly small than its distance to the central body A then the "gravitational force" will be "felt" by the satellite body B only after its acceleration towards or away from the central body A will be obstructed by either internal or external means.

Figure 11.6 shows the ratio of the spacetime to Newton's gravitational force F_{gs}/F_{gN} defined by Eqs. (11.6-3) and (11.6-4) as a function of the relative distance b_1 between the bodies A and B when velocity of the body B is equal to zero ($\gamma_V = 0$).

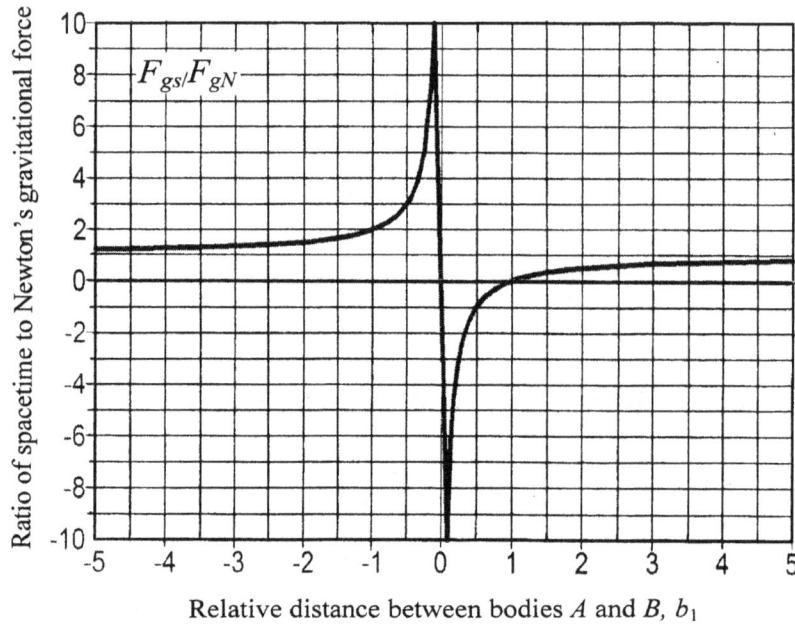

Figure 11.6. Ratio of the spacetime to Newton's gravitational force F_{gs}/F_{gN} as a function of the relative distance b_1 between bodies A and B.

11.7   Quantum States of Macro-Toryces of Celestial Bodies

Quantum states of macro-toryces of celestial bodies are described by the same quantization equations as the quantization equations of micro-toryces shown in Table 6.2. The quantization parameter z of a macro-toryx is expressed in Exhibit 11.7.

Exhibit 11.7. The quantization parameter z of a macro-toryx.

The quantization parameter z of a macro-toryx is expressed by the equation:

$$z = 2(k_b n \Lambda)^2 \qquad (11.7\text{-}1)$$

where k_b is the constant celestial body parameter.

Table 11.7.1. Parameters and status of planets in our solar system.

n	Planets	Measured distance to Sun r_{1m}, m	Relative measured dist. to Sun b_{1m}	Relative calculated dist. to Sun b_{1c}	Ratio b_{1m}/b_{1c}	Status of Sun's planets
1	Sun1	6.2378×10^{09}	8.4458×10^{6}	8.4458×10^{6}	1.0000	Former planets swallowed by the Sun
2	Sun 2	2.4951×10^{10}	3.3784×10^{7}	3.3784×10^{7}	1.0000	
3	**Mercury**	5.7909×10^{10}	7.8408×10^{7}	7.6014×10^{7}	**1.0315**	
4	**Venus**	1.0821×10^{11}	1.4651×10^{8}	1.3514×10^{8}	**1.0842**	**Currently existing**
5	**Earth**	1.4960×10^{11}	2.0255×10^{8}	2.1115×10^{8}	**0.9593**	
6	**Mars**	2.2794×10^{11}	3.0863×10^{8}	3.0406×10^{8}	**1.0150**	
7	Hungaria	3.0567×10^{11}	4.1387×10^{8}	4.1387×10^{8}	1.0000	Future planets that could be formed from asteroid belts
8	Phocaea	3.9923×10^{11}	5.4055×10^{8}	5.4055×10^{8}	1.0000	
9	Cybele	5.0527×10^{11}	6.8412×10^{8}	6.8412×10^{8}	1.0000	
10	Hilda	6.2382×10^{11}	8.4464×10^{8}	8.4464×10^{8}	1.0000	
11	**Jupiter**	7.7833×10^{11}	1.0538×10^{9}	1.0220×10^{9}	**1.0312**	**Currently existing**
12	Io	8.9826×10^{11}	1.2162×10^{9}	1.2162×10^{9}	1.0000	Former planets captured by the Jupiter and became its moons
13	Europe	1.0542×10^{12}	1.4273×10^{9}	1.4273×10^{9}	1.0000	
14	Ganymede	1.2227×10^{12}	1.6555×10^{9}	1.6555×10^{9}	1.0000	
15	**Saturn**	1.4270×10^{12}	1.9321×10^{9}	1.9004×10^{9}	**1.0167**	**Currently existing**
16	Tethys	1.5970×10^{12}	1.1622×10^{9}	1.1622×10^{9}	1.0000	Former planets captured by the Saturn and became its moons
17	Dione	1.1029×10^{12}	2.4410×10^{9}	2.4410×10^{9}	1.0000	
18	Rhea	2.0210×10^{12}	2.7364×10^{9}	2.7364×10^{9}	1.0000	
19	Titan	2.2518×10^{12}	3.0489×10^{9}	3.0489×10^{9}	1.0000	
20	Iapetus	2.4951×10^{12}	3.3783×10^{9}	3.3783×10^{9}	1.0000	
21	**Uranus**	2.8696×10^{12}	3.8853×10^{9}	3.7247×10^{9}	**1.0431**	**Currently existing**
22	Miranda	3.0190×10^{12}	4.0877×10^{9}	4.0877×10^{9}	1.0000	Former planets captured by the Uranus and became its moons
23	Ariel	3.2998×10^{12}	4.4678×10^{9}	4.4678×10^{9}	1.0000	
24	Umbriel	3.5929×10^{12}	4.8647×10^{9}	4.8647×10^{9}	1.0000	
25	Titania	3.8988×10^{12}	5.2789×10^{9}	5.2789×10^{9}	1.0000	
26	Oberon	4.2170×10^{12}	5.7097×10^{9}	5.7097×10^{9}	1.0000	
27	**Neptune**	4.4966×10^{11}	6.0883×10^{9}	6.0883×10^{9}	**0.9888**	**Currently existing**
28	Proteus	4.8904×10^{12}	6.6214×10^{9}	6.6214×10^{9}	1.0000	Former planets captured by the Neptune and became its moons
29	Triton	5.2460×10^{12}	7.1029×10^{9}	7.1029×10^{9}	1.0000	

Table 11.7.2. Parameters and status of the Jupiter's moons.

n	Jupiter's moons	Measured distance to Jupiter r_{1m}, m	Relative measured distance to Jupiter b_{1m}	Relative calculated distance to Jupiter b_{1c}	Ratio b_{1m}/b_{1c}	Status of Jupiter's moons
1	Jupiter 1	2.7115×10^7	3.8439×10^7	3.8439×10^7	1.0000	Former moons swallowed by the Jupiter
2	Jupiter 2	1.0846×10^8	1.5376×10^8	1.5376×10^8	1.0000	
3	Io	2.6200×10^8	3.7143×10^8	3.4595×10^8	1.0737	Currently existing moons
4	Europa	4.1690×10^8	5.9103×10^8	6.1502×10^8	0.9610	
5	Ganymede	6.6490×10^8	9.4261×10^8	9.6097×10^8	0.9809	
7	Callisto	1.1701×10^9	1.6588×10^9	1.8835×10^9	0.8807	

Table 11.7.3. Parameters and status of the Saturn's moons.

n	Saturn's moons	Measured distance to Saturn r_{1m}, m	Relative measured distance to Saturn b_{1m}	Relative calculated distance to Saturn b_{1c}	Ratio b_{1m}/b_{1c}	Status of Saturn's moons
1	Saturn 1	2.3974×10^7	1.1355×10^8	1.1355×10^8	1.0000	Former moons swallowed by the Saturn
2	Saturn 2	9.5890×10^7	4.5421×10^8	4.5421×10^8	1.0000	
3	Tethys	2.9462×10^8	1.3955×10^9	1.0220×10^9	1.3655	Currently existing moons
4	Dione	3.7740×10^8	1.7876×10^9	1.8168×10^9	0.9839	
5	Rhea	5.2711×10^8	2.4967×10^9	2.8388×10^9	0.8795	
7	Titan	1.2219×10^9	5.7877×10^9	5.5641×10^9	1.0402	
12	Iapetus	3.5608×10^9	1.6866×10^{10}	1.6352×10^{10}	1.0315	

Table 10.7.4. Parameters and status of the Uranus' moons.

N	Uranus' moons	Measured distance to Uranus r_{1m}, m	Relative measured distance to Uranus b_{1m}	Relative calculated distance to Uranus b_{1c}	Ratio b_{1m}/b_{1c}	Status of Uranus' moons
1	Uranus 1	1.2111×10^7	3.7538×10^8	3.7538×10^8	1.0000	Former moons swallowed by the Uranus
2	Uranus 2	4.8446×10^7	1.5015×10^9	1.5015×10^9	1.0000	
3	Miranda	1.2978×10^8	4.0225×10^9	3.3784×10^9	1.1907	Currently existing moons
4	Ariel	1.9102×10^8	5.9207×10^9	6.0061×10^9	0.9858	
5	Umbriel	2.6630×10^8	8.2540×10^9	9.3845×10^9	0.8795	
6	Titania	4.3591×10^8	1.3511×10^{10}	1.3514×10^{10}	0.9998	
7	Oberon	5.8352×10^8	1.8086×10^{10}	1.8394×10^{10}	0.9833	

Table 10.7.5. Parameters and status of the Neptun's moons.

n	Neptune's moons	Measured distance to Neptune r_{1m}, m	Relative measured distance to Neptune b_{1m}	Relative calculated distance to Neptune b_{1c}	Ratio b_{1m}/b_{1c}	Status of Neptune's moons
1	Neptune 1	1.4352×10^7	3.7538×10^8	3.7538×10^8	1.0000	Former moons swallowed by the Neptune
2	Neptune 2	4.8446×10^7	1.5015×10^9	1.5015×10^9	1.0000	
3	Proteus	1.1765×10^8	3.0772×10^9	3.3784×10^9	0.9108	Currently existing moons
5	Triton	3.5476×10^8	9.2791×10^9	9.3845×10^9	0.9888	

Table 11.7.6. Parameters of largest celestial bodies of our solar system.

Celestial body	k_b	m_A, kg	r_b, m	r_{0m}, m
Sun	15	1.98910×10^{30}	6.9550×10^8	738.566556
Jupiter	32	1.89973×10^{27}	7.1492×10^7	0.705383
Saturn	55	5.68598×10^{26}	6.0268×10^7	0.211124
Uranus	100	8.68910×10^{25}	2.5559×10^7	0.032263
Neptune	100	1.02966×10^{26}	2.4764×10^7	0.038232

Tables 11.7.1 – 11.7.5 show parameters of planets and moons in our solar system, and also their possible past, current and projected future locations. Table 11.7.6 shows the values of the celestial body parameters k_b, the body masses m_A, the equatorial body radii r_b and the mass eye radii r_{0m} for the largest bodies of our solar system. Notably, the celestial body parameters k_b is probably dependent on both density and distribution of mass inside the body.

11.8 Galaxy Law of Star Motion

Consider a star moving along a circular path with the radius r_s around a galaxy center with the orbital velocity V_s as shown in Fig. 11.8. In that case, the eye radius of the macro-toryx r_{0g} associated with the star depends on the galaxy mass located within a sphere with the radius equal to the star orbital radius r_s as defined in Exhibit 11.8.

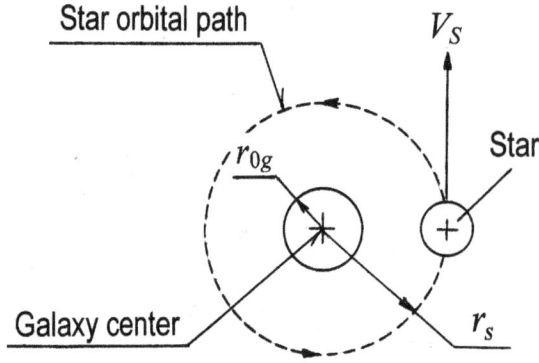

Figure 11.8. A star moving along a circular path with the radius r_s around a galaxy center with the orbital velocity V_s.

Exhibit 11.8. The eye radius of a macro-toryx r_{0g} associated with a star.

- The eye radius of a macro-toryx r_{0g} associated with a star is directly-proportional to the galaxy mass m_G located within a sphere with the radius equal to the star orbital radius r_s:

$$r_{0g} = \frac{m_G G}{2c^2} \qquad (11.8\text{-}1)$$

- When the star orbital radius r_s is much greater than the eye radius of macro-toryx r_{0g}, the ratio s of the galaxy mass m_G to the star orbital radius r_s is constant:

$$\frac{m_G}{r_s} = s = const. \qquad (11.8\text{-}2)$$

For the case when the star orbital radius r_s is much greater than the eye radius of macro-toryx r_{0g}, Eq. (11.4-1) reduces to the form representing the **galaxy law of star motion** in relative units:

$$\beta_s = \frac{V_s}{c} = \sqrt{\frac{2}{b_s}} \qquad (11.8\text{-}3)$$

where from Eqs. (11.4-3), (11.4-4) and (11.8-1) the relative star orbital radius b_s is equal to:

$$b_s = \frac{r_s}{r_{0g}} = \frac{2c^2}{sG} = const. \qquad (11.8\text{-}4)$$

Consequently, the galaxy law of star motion expressed by Eq. (11.8-3) reduces the form representing the spacetime law of star motion in absolute units::

$$V_s = \sqrt{\frac{m_G G}{r_s}} = \sqrt{sG} = const. \tag{11.8-5}$$

Thus, for the case when the star orbital radius r_s is much greater than the radius of the galaxy's black hole, the star velocity V_s is constant and does not depend on the star orbital radius r_s.

11.9 Outverted & Inverted Stars

Depending on the relationship between the outer radius r_b of a star body and the eye radius r_{0m} of a macro-toryx associated with a star, the stars can be divided into three kinds: *outverted stars*, *inverted stars* and *imaginary stars* as shown in Figure 11.9.

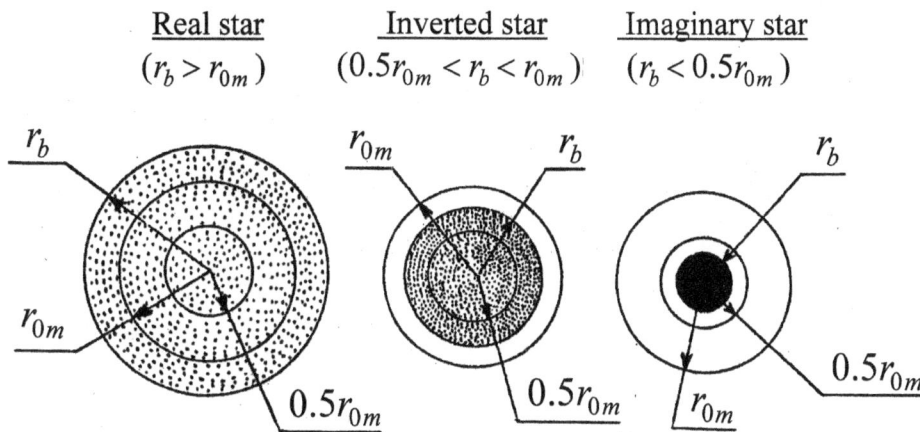

Figure 11.9. Three kinds of stars.

Outverted stars $(r_b \gg r_{0m})$ – In the outverted stars, the outer radius r_b of the star body is much greater than the eye radius r_{0m} of a macro-toryx associated with this star.

Marginally-outverted stars $(r_b > r_{0m})$ – In the marginally-outverted stars, the outer radius r_b of the star body is slightly greater than the eye radius r_{0m} of a macro-toryx associated with this star.

Inverted stars $(0.5r_b < r_b < r_{0m})$ – In the inverted stars, the outer radius r_b of a star body is greater than one half of the eye radius r_{0m}, but less than the eye radius r_{0m} of macro-toryx associated with this star.

Imaginary stars ($r_b < 0.5r_b$) – In the imaginary stars, the outer radius r_b of a star body is less than one half of the eye radius r_{0m} of a macro-toryx associated with this star.

The proposed theory considers the following examples of the stars:

- Our Sun is a typical example of an outverted star
- The *neutron stars* and *pulsars* are typical examples of the marginally-outverted stars
- The inverted and imaginary stars represent two phases of *black holes.*

Notes

<u>Notes</u>

PART 3

Abstract
Mathematics
of a Helyx

12. HELYX BASIC STRUCTURE & PARAMETERS

CONTENTS

12.1 Helyx Basic Structure

As shown in Fig. 12.1.1, the helyx basic structure is made up of a **leading string** O_1O_1 and a double-helical **trailing string**, both residing inside a **cylindrical boundary**. The leading string can be envisioned as a trace left by a point a propagating with the translational velocity $\tilde{V}_{1t}$ along a line O_1O_1 with the radius $\tilde{r}_1 \to \infty$. The double-helical trailing string can be envisioned as two traces left by two points, m and n, rotating around the point a along a circle with the radius $\tilde{r}_2$ with the rotational velocity $\tilde{V}_{2r}$, while the point a propagates along the line O_1O_1 with the translational velocity $\tilde{V}_{2t}$. The traces left by the moving points, m and n, form two helical spirals winding around the point a. In Fig. 12.1.1, $\tilde{\varphi}_1$ is the apex angle of helyx trailing string.

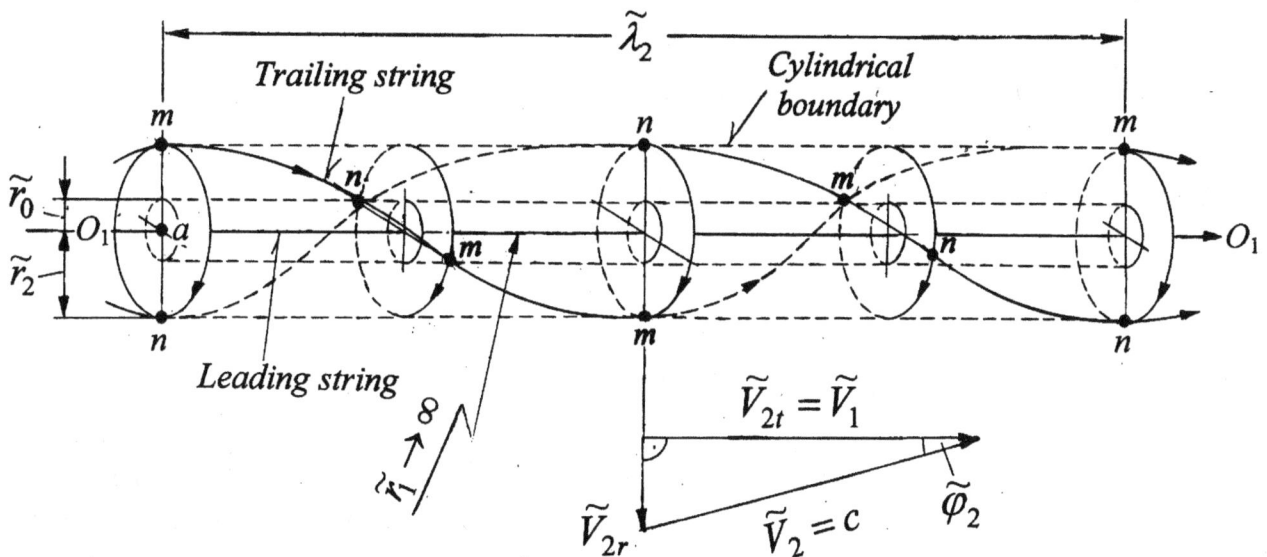

Figure 12.1.1. Helyx structure.

The leading string can be thought as a particular case of a helical spiral string in which the rotational velocity $\widetilde{V}_{1r}$ is equal to zero, while its translational velocity $\widetilde{V}_{1t}$ is equal to its spiral velocity $\widetilde{V}_1$, so

$$\widetilde{V}_{1r} = 0 \qquad (12.1\text{-}1)$$

$$\widetilde{V}_{1t} = \widetilde{V}_1 \qquad (12.1\text{-}2)$$

In each branch of the trailing string, the spiral velocity $\widetilde{V}_2$ is related to the translational velocity $\widetilde{V}_{2t}$ and the rotational velocity $\widetilde{V}_{2r}$ by the Pythagorean Theorem:

$$\widetilde{V}_2 = \sqrt{\widetilde{V}_{2t}^2 + \widetilde{V}_{2r}^2} \qquad (12.1\text{-}3)$$

The radius of helyx cylindrical boundary $\widetilde{r}$ is equal to:

$$\widetilde{r} = \widetilde{r}_2 \qquad (12.1\text{-}4)$$

Helyx spin - Helyx may have either up or down spins, with both of them defined by the right-hand rule as shown in Fig. 12.1.2. The helyx spin depends on the directions of the rotational velocities of helyx trailing string V_{2r}.

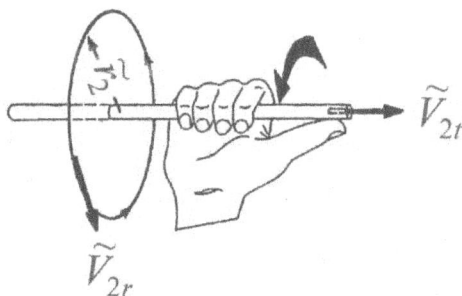

Figure 12.1.2. The up helyx spin.

Fine structure of helyx strings – When considering a fine structure of helyx strings, they appear as helical spirals with the radius $\widetilde{r}_s$ corresponding to the limits of spacetime equal to the Planck length l_p divided by 2π:

$$\widetilde{r}_s = \frac{l_p}{2\pi} \qquad (12.1\text{-}5)$$

12.2 Helyx Spacetime Parameters in Absolute Units

$\tilde{f}_0$ = helyx base frequency

$\tilde{f}_1$ = frequency of helyx leading string

$\tilde{f}_2$ = frequency of helyx trailing string

$\tilde{L}_1$ = spiral length of one winding of helyx leading string

$\tilde{L}_2$ = spiral length of one winding of helyx trailing string

$\tilde{r}$ = radius of helyx cylindrical boundary

$\tilde{r}_0$ = helyx eye radius

$\tilde{r}_1$ = radius of helyx leading string

$\tilde{r}_2$ = radius of helyx trailing string

$\tilde{T}_0$ = helyx base period

$\tilde{T}_1$ = period of helyx leading string

$\tilde{T}_2$ = period of helyx trailing string

$\tilde{V}_1$ = spiral velocity of helyx leading string

$\tilde{V}_{1r}$ = rotational velocity of helyx leading string

$\tilde{V}_{1t}$ = translational velocity of helyx leading string

$\tilde{V}_2$ = spiral velocity of helyx trailing string

$\tilde{V}_{2r}$ = rotational velocity of helyx trailing string

$\tilde{V}_{2t}$ = translational velocity of helyx trailing string

$\tilde{\lambda}_1$ = wavelength of helyx leading string

$\tilde{\lambda}_2$ = wavelength of helyx trailing string

$\tilde{\varphi}_2$ = apex angle of helyx trailing string.

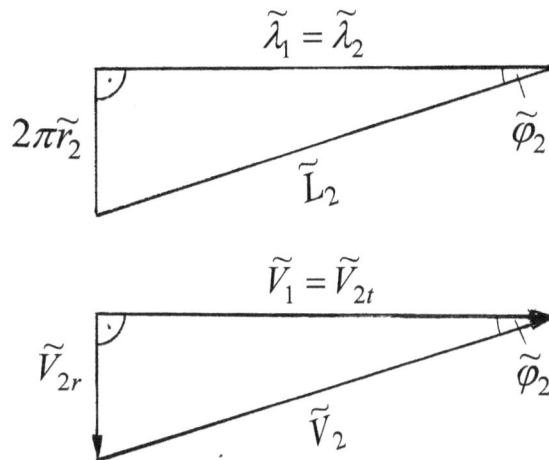

Figure 12.2. Helyx spacetime parameters in absolute units.

Figure 12.2 shows a hodograph of helyx spacetime parameters in absolute units. Symbols used for defining the helyx parameters are the same as those used for the toryx, except for the "wave" mark (tilde) over the symbols of the helyx parameters.

12.3 Helyx Spacetime Postulates in Absolute Units

The helyx spacetime postulates in absolute units are shown in Exhibit 12.3. This is a set of three fundamental equations limiting the degrees of several helyx parameters, making possible to establish relationships between all spacetime parameters of the helyx within the range of the spiral length of helyx trailing string $\widetilde{L}_2$ extending from positive to negative infinity.

Exhibit 12.3. Helyx spacetime postulates in absolute units.

- The radius of helyx eye $\widetilde{r}_0$ is equal to the radius of toryx eye r_0:

$$\widetilde{r}_0 = r_0 \qquad (12.3\text{-}1)$$

- The difference between the spiral length of helyx trailing string $\widetilde{L}_2$ and the wavelength of helyx trailing string $\widetilde{\lambda}_2$ is constant and equals to the helyx eye perimeter $2\pi\widetilde{r}_0$:

$$\widetilde{L}_2 - \widetilde{\lambda}_2 = 2\pi\widetilde{r}_0 = const. \qquad (12.3\text{-}2)$$

- The spiral velocity of helyx trailing string $\widetilde{V}_2$ is constant at each point of its spiral path:

$$\widetilde{V}_2 = \sqrt{\widetilde{V}_{2t}^2 + \widetilde{V}_{2r}^2} = c = const. \qquad (12.3\text{-}3)$$

The helyx base frequency $\widetilde{f}_0$ is equal to its parental toryx base frequency f_0, so:

$$\widetilde{f}_0 = f_0 = \frac{c}{2\pi\widetilde{r}_0} = \frac{c}{2\pi r_0} \qquad (12.3\text{-}4)$$

12.4 Helyx Spacetime Parameters in Relative Units

The helyx spacetime postulates can be simplified by expressing the helyx spacetime parameters in relative values in respect to the helyx eye radius $\tilde{r}_0$, the velocity of light c and the helyx base frequency $\tilde{f}_0$ as shown in Table 12.4.

Table 12.4. Helyx relative spacetime parameters.

Toryx relative parameters	Equations
Radius of helyx cylindrical boundary	$\tilde{b} = \tilde{r} / \tilde{r}_0$ (12.4-1)
Radius of helyx eye	$\tilde{b}_0 = 1$ (12.4-2)
Radius of helyx leading string	$\tilde{b}_1 = \tilde{r}_1 / \tilde{r}_0$ (12.4-3)
Radius of helyx trailing string	$\tilde{b}_2 = \tilde{r}_2 / \tilde{r}_0$ (12.4-4)
Length of helyx leading string	$\tilde{l}_1 = \tilde{L}_1 / 2\pi\tilde{r}_0$ (12.4-5)
Length of helyx trailing string	$\tilde{l}_2 = \tilde{L}_2 / 2\pi\tilde{r}_0$ (12.4-6)
Period of helyx leading string	$\tilde{t}_1 = \tilde{T}_1 / \tilde{f}_0$ (12.4-7)
Period of helyx trailing string	$\tilde{t}_2 = \tilde{T}_2 / \tilde{f}_0$ (12.4-8)
Spiral velocity of helyx leading string	$\tilde{\beta}_1 = \tilde{V}_1 / c$ (12.4-9)
Translational velocity of helyx leading string	$\tilde{\beta}_{1t} = \tilde{V}_{1t} / c$ (12.4-10)
Rotational velocity of helyx leading string	$\tilde{\beta}_{1r} = \tilde{V}_{1r} / c$ (12.4-11)
Spiral velocity of helyx trailing string	$\tilde{\beta}_2 = \tilde{V}_2 / c$ (12.4-12)
Translational velocity of helyx trailing string	$\tilde{\beta}_{2t} = \tilde{V}_{2t} / c$ (12.4-13)
Rotational velocity of helyx trailing string	$\tilde{\beta}_{2r} = \tilde{V}_{2r} / c$ (12.4-14)
Frequency of helyx leading string	$\tilde{\delta}_1 = \tilde{f}_1 / \tilde{f}_0$ (12.4-15)
Frequency of helyx trailing string	$\tilde{\delta}_2 = \tilde{f}_2 / \tilde{f}_0$ (12.4-16)
Wavelength of helyx leading string	$\tilde{\eta}_1 = \tilde{\lambda}_1 / 2\pi\tilde{r}_0$ (12.4-17)
Wavelength of helyx trailing string	$\tilde{\eta}_2 = \tilde{\lambda}_2 / 2\pi\tilde{r}_0$ (12.4-18)

12.5 Helyx Spacetime Postulates in Relative Units

Exhibit 12.5 shows three helyx spacetime postulates in relative units.

Exhibit 12.5. Helyx spacetime postulates in relative units.

- The relative radius of helyx eye $\tilde{b}_0$ is equal to the relative radius of toryx eye b_0 and is equal to 1:

$$\tilde{b}_0 = b_0 = 1 \qquad (12.5\text{-}1)$$

- The difference between the relative spiral length of helyx trailing string $\tilde{l}_2$ and the relative wavelength of helyx trailing string $\tilde{\eta}_2$ is equal to 1:

$$\tilde{l}_2 - \tilde{\eta}_2 = 1 \qquad (12.5\text{-}2)$$

- The relative spiral velocity of helyx trailing string $\tilde{\beta}_2$ is equal to 1 at each point of its spiral path:

$$\tilde{\beta}_2 = \sqrt{\tilde{\beta}_{2t}^2 + \tilde{\beta}_{2r}^2} = 1 \qquad (12.5\text{-}3)$$

Figure 12.5 shows the hodograph of helyx parameters in relative units.

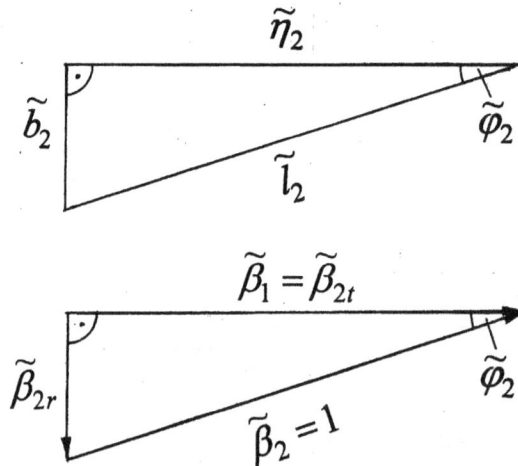

Figure 12.5. Helyx relative spacetime parameters in relative units.

12.6 Summary of Derived Helyx Spacetime Equations

Based on the basic helyx structure and also on the equations of helyx spacetime postulates, it is possible to derive equations for all other helyx spacetime parameters shown in Table 12.6.

Table 12.6. Spacetime parameters of helyx trailing string as the functions of its relative length $\tilde{l}_2$ and its apex angle $\tilde{\varphi}_2$.

Relative parameter	As a function of $\tilde{l}_2$ Eq.(a)	As a function of $\tilde{\varphi}_2$ Eq. (b)
Apex angle Eq.(12.6-1)	$\cos s\tilde{\varphi}_2 = \dfrac{\tilde{l}_2 - 1}{\tilde{l}_2}$	-
Length Eq.(12.6-2)	$\tilde{l}_2 = \dfrac{\tilde{L}_2}{2\pi \tilde{r}_0} = \dfrac{1}{\tilde{\delta}_2}$	$\tilde{l}_2 = \dfrac{\tilde{L}_2}{2\pi \tilde{r}_0} = \dfrac{1}{1 - \cos s\tilde{\varphi}_2}$
Wavelength Eq.(12.6-3)	$\tilde{\eta}_2 = \dfrac{\tilde{\lambda}_2}{2\pi \tilde{r}_0} = \tilde{l}_2 - 1$	$\tilde{\eta}_2 = \dfrac{\tilde{\lambda}_2}{2\pi \tilde{r}_0} = \dfrac{\cos s\tilde{\varphi}_2}{1 - \cos s\tilde{\varphi}_2}$
Radius Eq.(12.6-4)	$\tilde{b}_2 = \dfrac{\tilde{r}_2}{\tilde{r}_0} = \sqrt{2\tilde{l}_2 - 1}$	$\tilde{b}_2 = \dfrac{\tilde{r}_2}{\tilde{r}_0} = \sqrt{\dfrac{1 + \cos s\tilde{\varphi}_2}{1 - \cos s\tilde{\varphi}_2}}$
Cylindrical boundary Eq.(12.6-5)	$\tilde{b} = \dfrac{\tilde{r}}{\tilde{r}_0} = \sqrt{2\tilde{l}_2 - 1}$	$\tilde{b} = \dfrac{\tilde{r}}{\tilde{r}_0} = \sqrt{\dfrac{1 + \cos s\tilde{\varphi}_2}{1 - \cos s\tilde{\varphi}_2}}$
Translational velocity Eq.(12.6-6)	$\tilde{\beta}_{2t} = \dfrac{\tilde{V}_{2t}}{c} = \dfrac{\tilde{l}_2 - 1}{\tilde{l}_2}$	$\tilde{\beta}_{2t} = \dfrac{\tilde{V}_{2t}}{c} = \cos s\tilde{\varphi}_2$
Rotational velocity Eq.(12.6-7)	$\tilde{\beta}_{2r} = \dfrac{\tilde{V}_{2r}}{c} = \dfrac{\sqrt{2\tilde{l}_2 - 1}}{\tilde{l}_2}$	$\tilde{\beta}_{2r} = \dfrac{\tilde{V}_{2r}}{c} = \sin s\tilde{\varphi}_2$
Spiral velocity Eq.(12.6-8)	$\tilde{\beta}_2 = \dfrac{\tilde{V}_2}{c} = 1$	$\tilde{\beta}_2 = \dfrac{\tilde{V}_2}{c} = 1$
Frequency Eq.(12.6-9)	$\tilde{\delta}_2 = \dfrac{\tilde{f}_2}{\tilde{f}_0} = \dfrac{1}{\tilde{l}_2}$	$\tilde{\delta}_2 = \dfrac{\tilde{f}_2}{\tilde{f}_2} = 1 - \cos s\tilde{\varphi}_2$
Period Eq.(12.6-10)	$\tilde{t}_2 = \dfrac{\tilde{f}_0}{\tilde{f}_2} = \tilde{l}_2$	$\tilde{t}_2 = \dfrac{\tilde{f}_2}{\tilde{f}_2} = \dfrac{1}{1 - \cos s\tilde{\varphi}_2}$

The toryx trigonometric function $\cos s\widetilde{\varphi}_2$ relates to the elementary trigonometric function $\cos s\varphi_2$ as follows:

$$\cos s\widetilde{\varphi}_2 = \cos\varphi_2 \quad (0 < \varphi_2 < 180^0) \tag{12.6-11}$$

$$\cos s\widetilde{\varphi}_2 = 1/\cos\varphi_2 \quad (180^0 < \varphi_2 < 360^0) \tag{12.6-12}$$

13. FEATURES OF ABSTRACT MATHEMATICS OF A HELYX

CONTENTS

13.1 Infinity versus Elementary Zero
13.2 Helyx Spacetime Trigonometry
13.3 Helyx Number Lines.

Equations describing the helyx spacetime parameters are mostly based on elementary math commonly taught in high schools. However, to satisfy the helyx spacetime postulates, it is necessary to modify several aspects of elementary math, including the definitions of zero, number line and elementary trigonometric functions. Also, unlike the elementary math that deals with stationary spiral elements, the helyx math considers the spiral elements in motion.

13.1 Infinility versus Elementary Zero

Conventionally, we use the elementary zero (0) in two ways. Firstly, we use it for counting of non-divisible entities. In an elementary number line (Fig. 13.1.1) it appears as an integer immediately preceding number one (1).

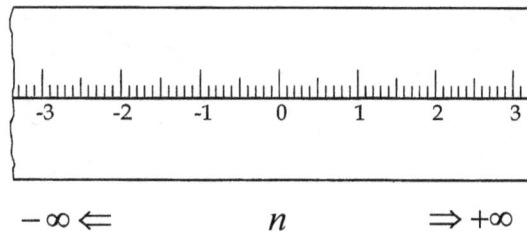

$$-\infty \Leftarrow \qquad n \qquad \Rightarrow +\infty$$

Figure 13.1.1. Elementary number line.

Secondly, we use zero to represent the absolute absence of any quantity and quality. Mathematically, the elementary zero (0) is equal to a ratio of one (1) to infinity (∞). The helyx math clearly separates two applications of zero described above. The zero is still considered as an integer for counting of non-divisible entities and still retains its old symbol (0). But, in application to the spacetime entities the zero is replaced with a quantity that is infinitely approaching to it. This quantity is called *infinility*, from the "infinite nil." (Notably, the term infinility is used in the helyx math instead of the known math term *infinitesimal*). In the helyx math, both infinity and infinility can be positive, negative, real and imaginary as shown below.

$$\text{Real infinility: } \pm 0 = \frac{1}{\pm \infty}; \quad \text{Imaginary infinility: } \pm 0i = \frac{1}{\pm \infty i}$$

$$\text{Real infinity: } \pm \infty = \frac{1}{\pm 0}; \quad \text{Imaginary infinity: } \pm \infty i = \frac{1}{\pm 0i}$$

Figure 13.1.2 shows symbolically positive and negative infinities ($\pm\infty$) and also positive and negative infinility (± 0) as equal counterparts in respect to the positive and negative unities (± 1).

Figure 13.1.2. Infinity ($\pm\infty$), infinility (±0) and unity (±1) .

13.2 Helyx Spacetime Trigonometry

Definitions of elementary trigonometric functions are based on transformations of a right triangle as a function of the non-right angle $\widetilde{\varphi}_2$ (Fig. 13.2.1):

$$\cos\varphi_2 = x \quad (0^0 < \varphi_2 < 360^0) \qquad (13.2\text{-}1)$$

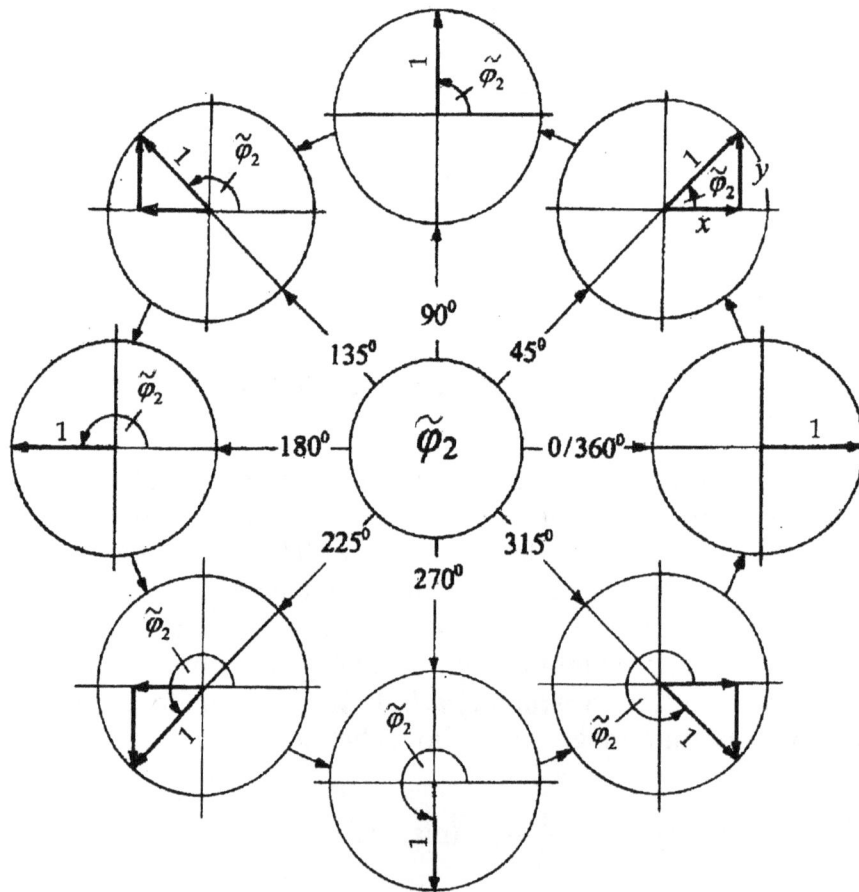

Figure 13.2.1. Transformations of a right triangle in elementary trigonometry.

The main features of the transformations shown in Fig. 13.2.1 are:

- When the length of the hypotenuse of the triangles is equal to 1, the ranges of the lengths of its sides x and y are between 1 and -1.
- The triangles located in two left quadrants are the mirror images of the triangles located in two right quadrants.
- The triangles located in two bottom quadrants are the mirror images of the triangles located at two top quadrants.

In the spacetime trigonometry, the transformations of the right triangle are partially modified to satisfy the helyx spacetime postulates. Consequently, the helyx spacetime trigonometric function $\cos s\widetilde{\varphi}_2$ relates to the elementary trigonometric function $\cos\widetilde{\varphi}_2$ as follows:

$$\cos s\widetilde{\varphi}_2 = \cos\widetilde{\varphi}_2 \quad (0 < \widetilde{\varphi}_2 < 180^0) \tag{13.2-2}$$

$$\cos s\widetilde{\varphi}_2 = 1/\cos\widetilde{\varphi}_2 \quad (180^0 < \widetilde{\varphi}_2 < 360^0) \tag{13.2-3}$$

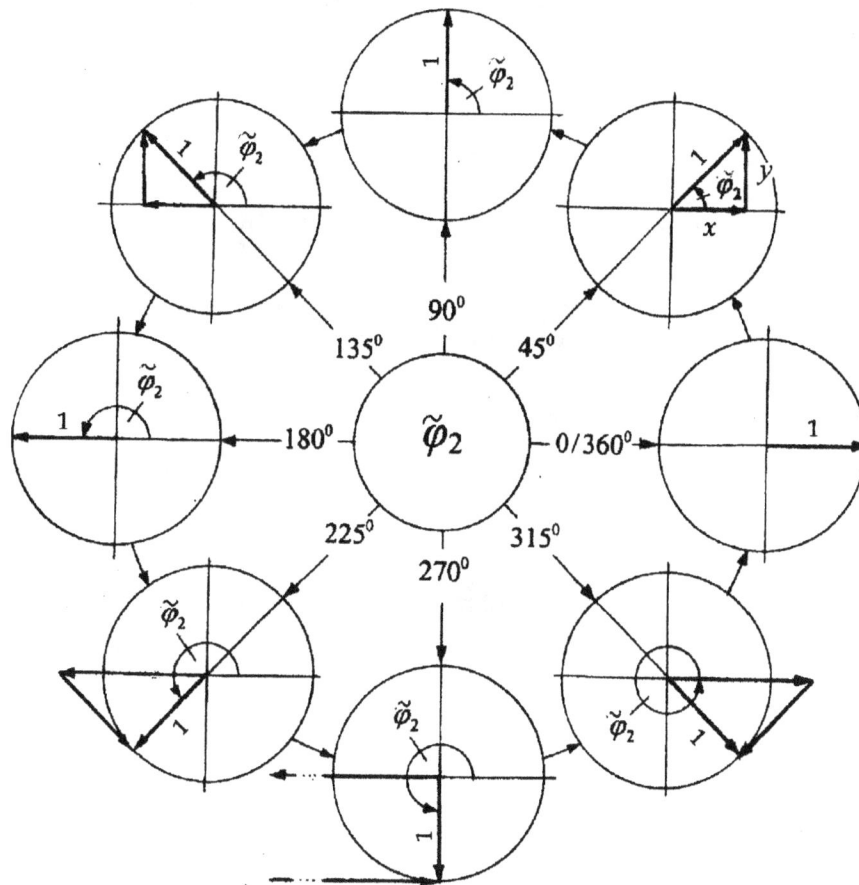

Figure 13.2.2. Transformations of a right triangle in the spacetime trigonometry.

The main features of the transformations shown in Fig. 13.2.2 are:

- When the angle $\widetilde{\varphi}_2$ is between 0 and 180^0, the right triangles are the same as in elementary trigonometry. Thus, within this range of the angle $\widetilde{\varphi}_2$ the elementary and spacetime trigonometry are based on the same principle.
- When the angle $\widetilde{\varphi}_2$ is between 180 and 360^0, the right triangle becomes **outverted**. Consequently, the length of its horizontal side x becomes greater than 1, while the length of the other side y is expressed with imaginary numbers.
- When the angle $\widetilde{\varphi}_2$ approaches 270^0 from the angle smaller than 270^0, the length of its horizontal side x approaches real positive infinity $(+\infty)$, while the length of the other side y approaches imaginary positive infinity $(+\infty i)$.
- When the angle $\widetilde{\varphi}_2$ approaches 270^0 from the angle greater than 270^0, the length of its horizontal side x approaches real negative infinity $(-\infty)$, while the length of the other side y approaches imaginary negative infinity $(-\infty i)$.
- When the angle $\widetilde{\varphi}_2$ approaches 360^0 from the angle smaller than 360^0, the length of its horizontal side x approaches 1, while the length of the other imaginary side y approaches imaginary negative infinility $(-0i)$.

13.3 Helyx Number Lines

We consider below two kinds of helyx number lines that are directly related to the helyx parameters:

- *Helyx vorticity* $\widetilde{V}$ number line
- *Helyx reality* $\widetilde{R}$ number line.

Both kinds of number lines are presented below in the forms of circular diagrams in which the numbers $\widetilde{V}$ and $\widetilde{R}$ are expressed as functions of the steepness angle of helyx trailing string $\widetilde{\varphi}_2$.

Helyx vorticity $\widetilde{V}$ **number line** - In the helyx vorticity $\widetilde{V}$ number line (Fig. 13.3.1), the real numbers $\widetilde{V}$ are equal to the ratio of the wavelength of helyx trailing string $\widetilde{\lambda}_2$ to the spiral length of helyx trailing string radius $\widetilde{L}_2$ with an opposite sign. These numbers are extended clockwise along a circle from the real positive infinity $(+\infty)$ to the real negative infinity $(-\infty)$ as a function of the apex angle of trailing string $\widetilde{\varphi}_2$.

$$\widetilde{V} = -\frac{\widetilde{\lambda}_2}{\widetilde{L}_2} = -\cos s\widetilde{\varphi}_2 \qquad (13.3\text{-}1)$$

The helyx vorticity $\widetilde{V}$ number line is divided into two domains, the $\widetilde{V}$ *infinility domain* and the $\widetilde{V}$ *infinity domain*, occupying equal sectors of the circular number line.

- The $\widetilde{V}$ infinility domain occupies two top quadrants; it contains the values of $\widetilde{V}$ extending clockwise from the real positive unity $(+1)$ and passing through infinility (± 0) to the real negative unity (-1).

- The $\widetilde{V}$ infinity domain resides in two bottom quadrants; it contains the values of $\widetilde{V}$ extending counterclockwise from the real positive unity $(+1)$ and passing through infinity $(\pm\infty)$ to the real negative unity (-1).

In the helyx vorticity $\widetilde{V}$ number line, the real positive infinility $(+0)$ merges with real negative infinility (-0) at $\widetilde{\varphi}_2 = 90^0$, while real negative infinity $(-\infty)$ merges with real positive infinity $(+\infty)$ at $\widetilde{\varphi}_2 = 270^0$.

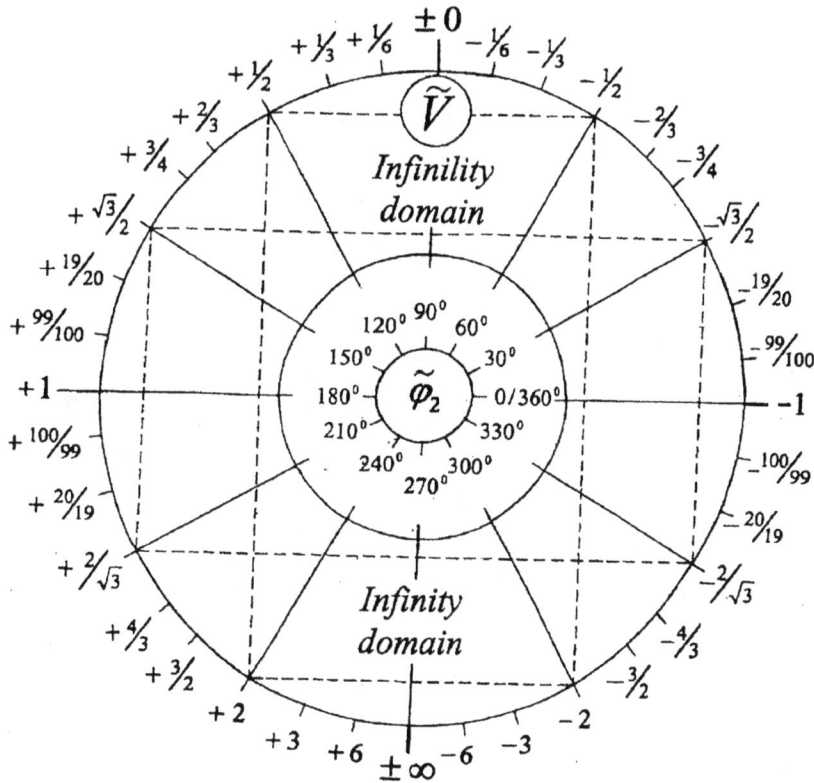

Figure 13.3.1. The helyx vorticity $\widetilde{V}$ number line.

There are two kinds of symmetries between the numbers $\widetilde{V}$ that belong to the four quadrants of circular diagram, the *inverse $\widetilde{V}$-symmetry* and the *reverse $\widetilde{V}$-symmetry*.

- In the inverse $\widetilde{V}$-symmetry, the magnitudes of the numbers $\widetilde{V}$ located in the top quadrants are inversed (reciprocated) in respect to the magnitudes of the numbers $\widetilde{V}$ located in the bottom quadrants.
- In the reverse $\widetilde{V}$-symmetry, the numbers $\widetilde{V}$ located in the right quadrants and the left quadrants have the same magnitudes but reversed signs.

The helyx vorticity $\widetilde{V}$ number line (Fig. 13.3.1) relates to the helyx parameters by the equation:

$$\widetilde{V} = -\frac{\widetilde{\lambda}_2}{\widetilde{L}_2} = -\frac{\widetilde{\eta}_2}{\widetilde{l}_2} = -\frac{\widetilde{l}_2 - 1}{\widetilde{l}_2} = \widetilde{\delta}_2 - 1 = -\widetilde{\beta}_{2t} = -\cos s\widetilde{\varphi}_2 \qquad (13.3\text{-}2)$$

Helyx reality $\widetilde{R}$ number line - In the helyx reality $\widetilde{R}$ number line (Fig. 13.3.2), the real and imaginary numbers $\widetilde{R}$ are equal to the ratio of helyx trailing string $\widetilde{r}_2$ to the radius of helyx eye $\widetilde{r}_0$.

$$\widetilde{R} = \frac{\widetilde{r}_2}{\widetilde{r}_0} = \sqrt{\frac{1 + \cos s\widetilde{\varphi}_2}{1 - \cos s\widetilde{\varphi}_2}} \qquad (13.3\text{-}3)$$

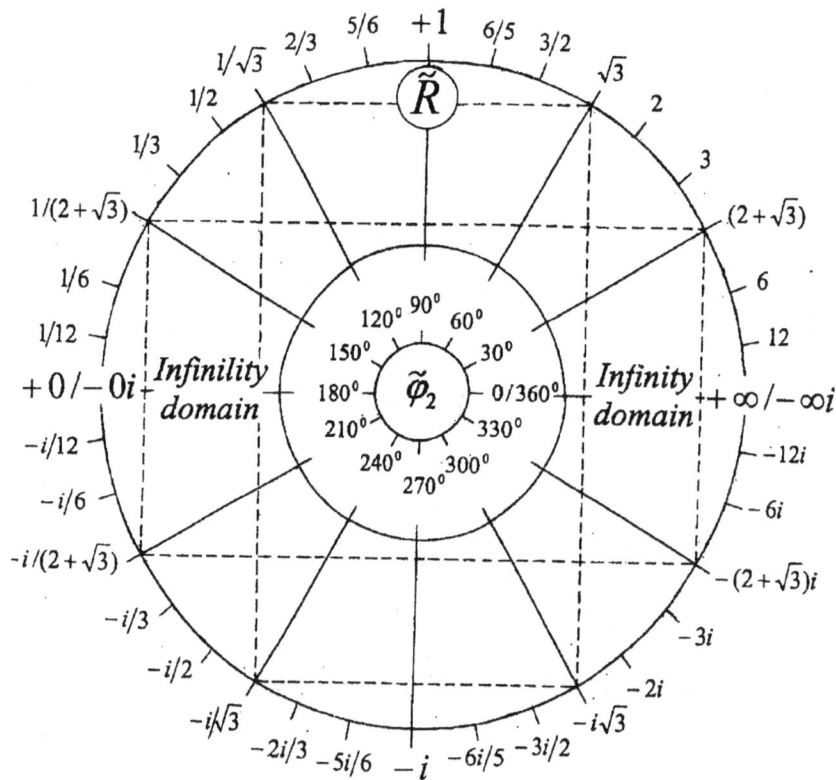

Figure 13.3.2. The helyx reality $\widetilde{R}$ number line.

These numbers are extended counterclockwise along a circle from the real positive infinity $(+\infty)$ to the imaginary negative infinity $(-\infty i)$ as a function of the steepness angle of trailing string $\widetilde{\varphi}_2$.

The helyx reality $\widetilde{R}$ number line is divided into two domains, the $\widetilde{R}$ *infinility domain* and the $\widetilde{R}$ *infinity domain,* occupying equal sectors of the circular number line.

- The $\widetilde{R}$ infinility domain occupies two left quadrants; it contains the values of $\widetilde{R}$ extending counterclockwise from the real positive unity $(+1)$ and passing through infinility $(+0/0i)$ to the imaginary negative unity $(-i)$.
- The $\widetilde{R}$ infinity domain resides in two right quadrants; it contains the values of $\widetilde{R}$ extending clockwise from the real positive unity $(+1)$ and passing through real positive and imaginary negative infinities $(+\infty/-\infty i)$ to the imaginary negative unity $(-i)$.

In the helyx reality $\widetilde{R}$ number line, the real positive infinility $(+0)$ merges with the imaginary negative infinility $(-0i)$ at $\widetilde{\varphi}_2 = 180^0$, while real positive infinity $(+\infty)$ merges with imaginary negative infinity $(-\infty i)$ at $\widetilde{\varphi}_2 = 360^0$.

There are two kinds of symmetries between the numbers $\widetilde{R}$ that belong to the four quadrants of circular diagram, the *inverse $\widetilde{R}$-symmetry* and the *reverse reality $\widetilde{R}$-symmetry*.

- In the inverse $\widetilde{R}$-symmetry, the magnitudes of the numbers $\widetilde{R}$ located in the left quadrants are inversed (reciprocated) in respect to the magnitudes of the numbers $\widetilde{R}$ located in the right quadrants.
- In the reverse reality $\widetilde{R}$-symmetry, the numbers $\widetilde{R}$ located in the top quadrants are real positive, while these numbers in the bottom quadrants are imaginary negative.

The numbers $\widetilde{R}$ the helyx reality $\widetilde{R}$ number line (Fig. 13.3.2) relate to the helyx parameters by the equation:

$$\widetilde{R} = \frac{\widetilde{r}_2}{\widetilde{r}_0} = \widetilde{b}_2 = \widetilde{l}_2 \beta_{2r} = \frac{\beta_{2r}}{\widetilde{\delta}_2} = \sqrt{\frac{1+\cos s\varphi_2}{1-\cos s\varphi_2}} \qquad (13.3\text{-}4)$$

Figure 13.3.3 shows the application of the spiral spacetime math for the calculation of relative velocities of helyx trailing string as its steepness angle $\widetilde{\varphi}_2$ increases from 0 to 360^0. In each right triangle of velocities of trailing string one side represents the relative translational velocity $\widetilde{\beta}_{2t}$ and the other side the relative rotational velocity $\widetilde{\beta}_{2r}$, while its hypotenuse represents the relative spiral velocity $\widetilde{\beta}_2 = 1$. There is a clear similarity between the transformations of the velocities of the helyx trailing string shown in Figure 13.3.3 and the transformations of a right triangle corresponding to the helyx trigonometry shown in Figure 13.2.2.

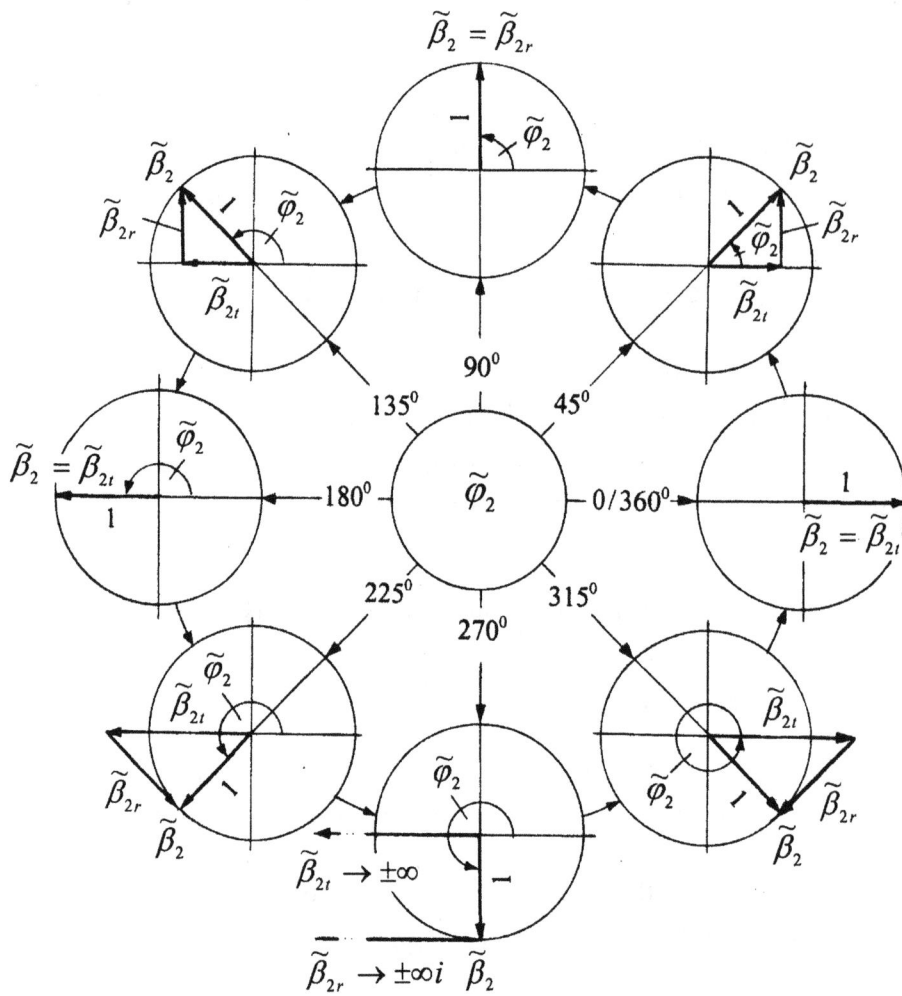

Figure 13.3.3. Transformations of right triangle representing vectors of the relative velocities of helyx trailing string $\tilde{\beta}_2$, $\tilde{\beta}_{2t}$ and $\tilde{\beta}_{2r}$.

Notes

Notes

14. CLASSIFICATION OF HELYCES

14.1 Main Groups & Subgroups of Helyces
14.2 Vorticities & Realities of Adjacent Helyces
14.3 Self-Polarized Helyces.

14.1 Main Groups & Subgroups of Helyces

Helyces are divided into four main groups and eight subgroups according to their vorticity $\widetilde{V}$ and reality $\widetilde{R}$ as shown in Tables 14.1.1, 14.1.2 and Figs. 14.1.1, 1.4.2.

Table 14.1.1. Realities $\widetilde{R}$ and vorticities $\widetilde{V}$ of helyces of main groups.

Helyx name	$\widetilde{\varphi}_2$	$\widetilde{R}$	$\widetilde{V}$
Real negative	0^0 - 90^0	Real	$(-)$
Real positive	90^0 - 180^0	Real	$(+)$
Imaginary positive	180^0 - 270^0	Imaginary	$(+)$
Imaginary negative	270^0 - 360^0	Imaginary	$(-)$

Real negative helyces $(0^0 < \widetilde{\varphi}_2 < 90^0)$ – The real negative helyces are located in the top right quadrants of the circular diagrams. In these helyces, the helyx reality $\widetilde{R}$ is expressed with real numbers and the helyx vorticity $\widetilde{V}$ with negative numbers.

Real positive helyces $(90^0 < \widetilde{\varphi}_2 < 180^0)$ – The real positive helyces are located in the top left quadrants of the circular diagrams. In these helyces, the helyx reality $\widetilde{R}$ is expressed with real numbers and the helyx vorticity $\widetilde{V}$ with positive numbers.

Imaginary positive helyces $(180^0 < \widetilde{\varphi}_2 < 270^0)$ – The imaginary positive helyces are located in the bottom left quadrants of the circular diagrams. In these helyces, the helyx reality $\widetilde{R}$ is expressed with imaginary numbers and the helyx vorticity $\widetilde{V}$ with positive numbers.

Imaginary negative helyces $(270^0 < \widetilde{\varphi}_2 < 360^0)$ – The imaginary negative helyces are located in the bottom right quadrants of the circular diagrams. In these helyces, the helyx reality $\widetilde{R}$ is expressed with imaginary numbers and the helyx vorticity $\widetilde{V}$ with negative numbers.

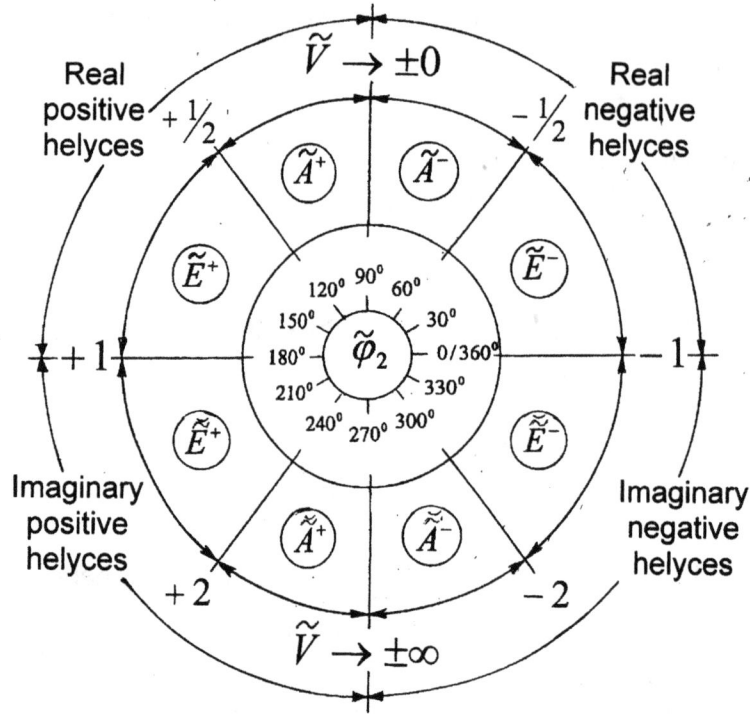

Figure 14.1.1. Main groups and subgroups of helyces as a function of helyx vorticity $\widetilde{V}$.

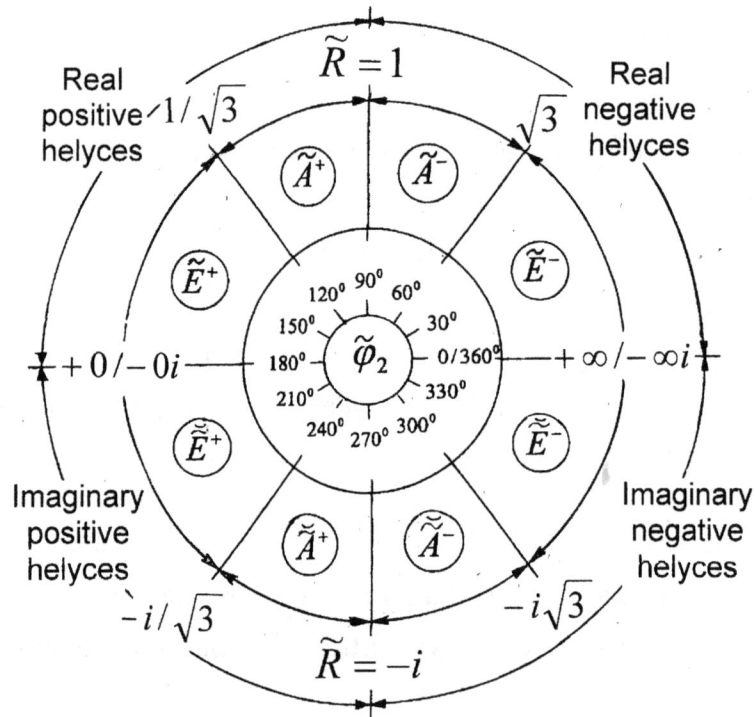

Figure 14.1.2. Main groups and subgroups of helyces as a function of helyx reality $\widetilde{R}$.

Within each main group, the helyces are further divided into two **subgroups** as shown in Table 14.1.2.

Table 14.1.2. Helyces of main groups and subgroups.

Helyces of main groups	Helyces of subgroups		Helyces of main groups	Helyces of subgroups	
Real negative	$\widetilde{E}^{-}$	$0^{0} < \widetilde{\varphi}_2 < 60^{0}$	Imaginary positive	$\widetilde{\widetilde{E}}^{+}$	$180^{0} < \widetilde{\varphi}_2 < 240^{0}$
	$\widetilde{A}^{-}$	$60^{0} < \widetilde{\varphi}_2 < 90^{0}$		$\widetilde{\widetilde{A}}^{+}$	$240^{0} < \widetilde{\varphi}_2 < 270^{0}$
Real positive	$\widetilde{A}^{+}$	$90^{0} < \widetilde{\varphi}_2 < 120^{0}$	Imaginary negative	$\widetilde{\widetilde{A}}^{-}$	$270^{0} < \widetilde{\varphi}_2 < 300^{0}$
	$\widetilde{E}^{+}$	$120^{0} < \widetilde{\varphi}_2 < 180^{0}$		$\widetilde{\widetilde{E}}^{-}$	$300^{0} < \widetilde{\varphi}_2 < 360^{0}$

14.2 Vorticities & Realities of Adjacent Helyces

Main groups of helyces - The vorticities $\widetilde{V}$ and the realities $\widetilde{R}$ of helyces of adjacent main groups are symmetrically related as shown in Table 14.2.1.

Table 14.2.1. Symmetrical relationships between the vorticities $\widetilde{V}$ and the realities $\widetilde{R}$ of polarized helyces of adjacent main groups.

Helyces of main groups		Eqs. (a)	Eqs. (b)
Reality-polarized negative helyces	$\widetilde{\widetilde{E}}^{-}$ & $\widetilde{E}^{-}$ Eq. (14.2-1)	$\widetilde{\widetilde{V}}_E^{-} = 1/\widetilde{V}_E^{-}$	$\widetilde{\widetilde{R}}_E^{-} = \pm i \widetilde{R}_E^{-}$
Reality-polarized positive helyces	$\widetilde{\widetilde{E}}^{+}$ & $\widetilde{E}^{+}$ Eq. (14.2-2)	$\widetilde{\widetilde{V}}_E^{+} = 1/\widetilde{V}_E^{+}$	$\widetilde{\widetilde{R}}_E^{+} = \pm i \widetilde{R}_E^{+}$
Vorticity-polarized real helyces	$\widetilde{A}^{+}$ & $\widetilde{A}^{-}$ Eq. (14.2-3)	$\widetilde{V}_A^{+} = -\widetilde{V}_A^{-}$	$\widetilde{R}_A^{+} = 1/\widetilde{R}_A^{-}$
Vorticity-polarized Imaginary helyces	$\widetilde{\widetilde{A}}^{+}$ & $\widetilde{\widetilde{A}}^{-}$ Eq. (14.2-4)	$\widetilde{\widetilde{V}}_A^{+} = -\widetilde{\widetilde{V}}_A^{-}$	$\widetilde{\widetilde{R}}_A^{+} = 1/\widetilde{\widetilde{R}}_A^{-}$

Subgroups of helyces - Table 14.2.2 summarizes the relationships between vorticities of helyces of subgroups.

Table 14.2.2. Relationships between vorticities of helyces of subgroups.

Helyces of subgroups		Equations
Real negative helyces	$\widetilde{A}^-$ & $\widetilde{E}^-$	$\widetilde{V}_A^- + \widetilde{V}_E^- = -1$ (14.2-5)
Real Positive helyces	$\widetilde{A}^+$ & $\widetilde{E}^+$	$\widetilde{V}_A^+ + \widetilde{V}_E^+ = +1$ (14.2-6)
Imaginary Positive helyces	$\widetilde{\widetilde{A}}^+$ & $\widetilde{\widetilde{E}}^+$	$\dfrac{1}{\widetilde{V}_A^+} + \dfrac{1}{\widetilde{V}_E^+} = +1$ (14.2-7)
Imaginary negative helyces	$\widetilde{\widetilde{A}}^-$ & $\widetilde{\widetilde{E}}^-$	$\dfrac{1}{\widetilde{V}_A^-} + \dfrac{1}{\widetilde{V}_E^-} = -1$ (14.2-8)

14.3 Self-Polarized Helyces

In the self-polarized helyces, both the relative translational velocity $\widetilde{\beta}_{2t}$ and the relative rotational velocity $\widetilde{\beta}_{2r}$ of trailing string are superluminal, while the relative translational velocity $\widetilde{\beta}_{2t}$ and the relative rotational velocity $\widetilde{\beta}_{2r}$ is imaginary. Table 14.3 shows ranges of parameters of self-polarized helyces.

Table 14.3. Ranges of parameters of self-polarized helyces.

Relative parameters of trailing string		Apex angle of trailing string $\widetilde{\varphi}_2$		
		180^0	270^0	$0^0/360^0$
Length	$\widetilde{l}_2$	$+0.5$	± 0	$-\infty$
Wavelength	$\widetilde{\eta}_2$	-0.5	-1.0	$-\infty$
Radius	$\widetilde{b}_2$	$-0i$	$-1.0i$	$-\infty i$
Translational velocity	$\widetilde{\beta}_{2t}$	-1.0	$\mp\infty$	$+1.0$
Rotational velocity	$\widetilde{\beta}_{2r}$	$-0i$	$-\infty i$	$-0i$
Frequency	$\widetilde{\delta}_2$	$+2.0$	$\pm\infty$	-0

Notes

Notes

15. TRENDS & INVERSIONS OF HELYX PARAMETERS

15.1 Helyx Spacetime Trigonometry

Helyx spacetime parameters change significantly as the length of its trailing string changes from positive to negative infinity. Mathematics describing the helyx spacetime parameters is similar to the mathematics applied to a toryx.

Trigonometry used for helyx is the same as the toryx spacetime trigonometry, except for one difference. In the toryx, trigonometric functions are related to the steepness angle of the toryx trailing string $\widetilde{\varphi}_2$, while in the helyx they are related to the apex angle of the helyx trailing string $\widetilde{\varphi}_2$. Thus, for the helyx, the relationship between the helyx spacetime trigonometric function $\cos s\widetilde{\varphi}_2$ and the relative length of helyx trailing string $\widetilde{l}_2$ is given by the equation:

$$\cos s\widetilde{\varphi}_2 = \frac{\widetilde{l}_2 - 1}{\widetilde{l}_2} \qquad (0^0 < \widetilde{\varphi}_2 < 360^0) \qquad (15.1\text{-}1)$$

Table 15.1 shows the relationship between spacetime and elementary trigonometric functions in application to the helyx.

Table 15.1. Relationship between spacetime and elementary trigonometric functions in application to helyx.

Spacetime trigonometry	Elementary trigonometry	
$(0^0 < \widetilde{\varphi}_2 < 360^0)$	$(0^0 < \widetilde{\varphi}_2 < 180^0)$	$(180^0 < \widetilde{\varphi}_2 < 360^0)$
$\cos s\widetilde{\varphi}_2$	$\cos\varphi_2$	$\sec\varphi_2$
$\sin s\widetilde{\varphi}_2$	$\sin\varphi_2$	$i\tan\varphi_2$

Sown below are the graphs expressing the spacetime parameters of helyx trailing string as functions of the apex angle of trailing string $\widetilde{\varphi}_2$.

15.2 Relative Wavelength & Spiral Length of Trailing String

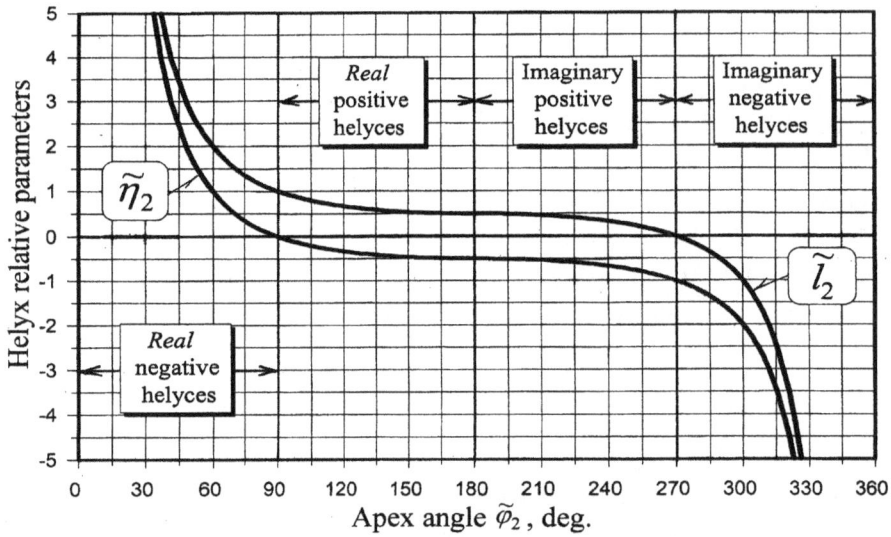

$\widetilde{\varphi}_2$	$360/0^0$	90^0	180^0	270^0	$\widetilde{\varphi}_2$	$360/0^0$	90^0	180^0	270^0
$\widetilde{\eta}_2$	$-\infty/+\infty$	$+0/-0$	$-\frac{1}{2}$	-1	$\widetilde{l}_2$	$-\infty/+\infty$	1	$+\frac{1}{2}$	$+0/-0$

Figure 15.2. Relative wavelength $\widetilde{\eta}_2$ and spiral length $\widetilde{l}_2$ of trailing string.

15.3 Relative Radius of Trailing String

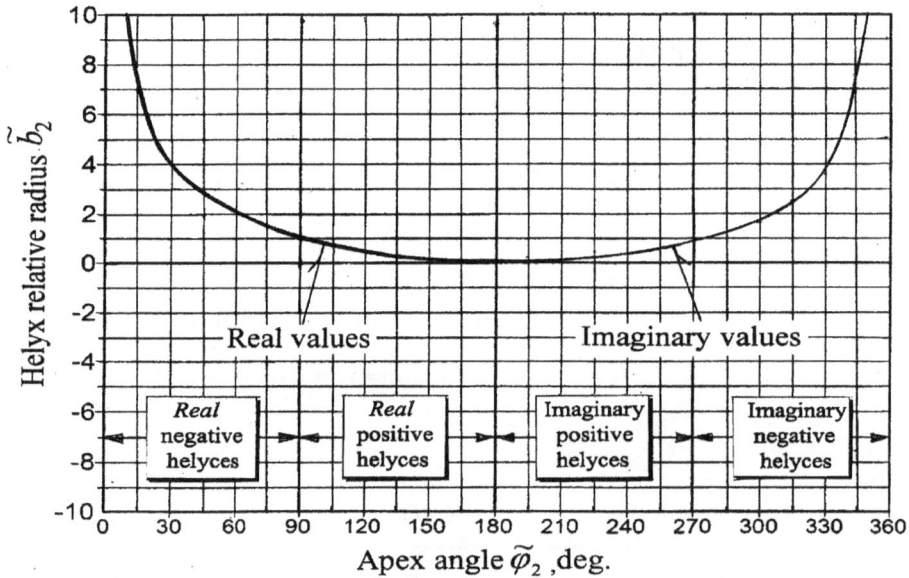

$\widetilde{\varphi}_2$	$360/0^0$	90^0	180^0	270^0
$\widetilde{b}_2$	$-\infty i/+\infty$	1.0	$+0/+0i$	$+1.0i$

Figure 15.3. Relative radius of trailing string $\widetilde{b}_2$.

15.4 Relative Translational Velocity of Trailing String

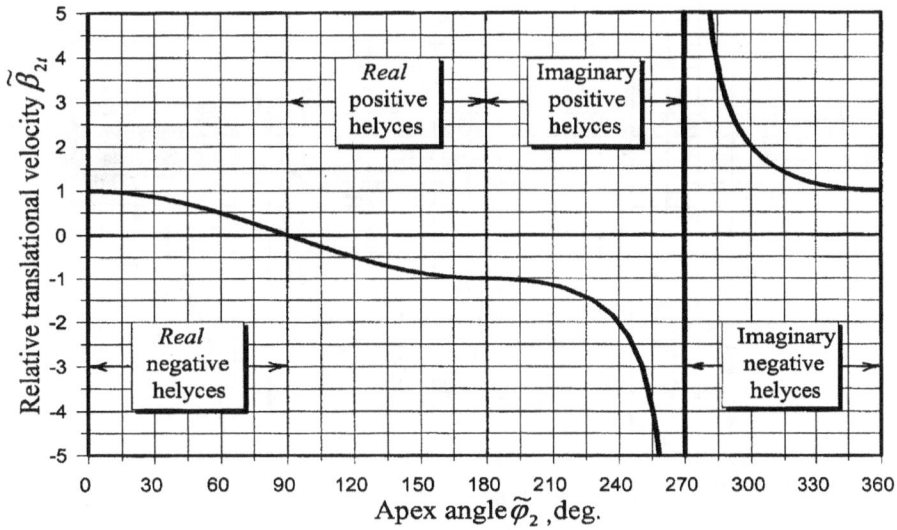

$\widetilde{\varphi}_2$	$360/0^0$	90^0	180^0	270^0
$\widetilde{\beta}_{2t}$	$+1$	$+0/-0$	-1	$-\infty/+\infty$

Figure 15.4. Relative translational velocity of trailing strings $\widetilde{\beta}_{2t}$.

15.5 Relative Rotational Velocity of Trailing String

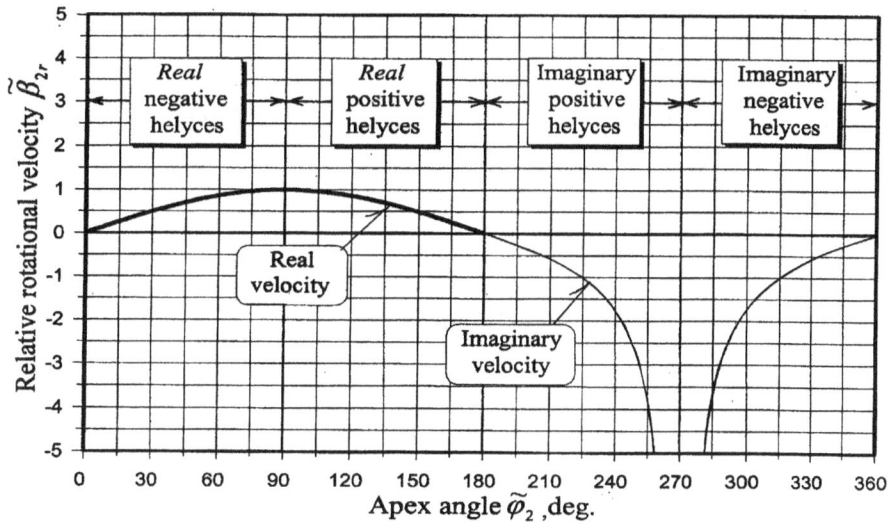

$\widetilde{\varphi}_2$	$360/0^0$	90^0	180^0	270^0
$\widetilde{\beta}_{2r}$	$-0i/+0$	1.0	$+0/-0i$	$-\infty i$

Figure 15.5. Relative rotational velocity of trailing strings $\widetilde{\beta}_{2r}$.

15.6 Relative Frequency of Trailing String

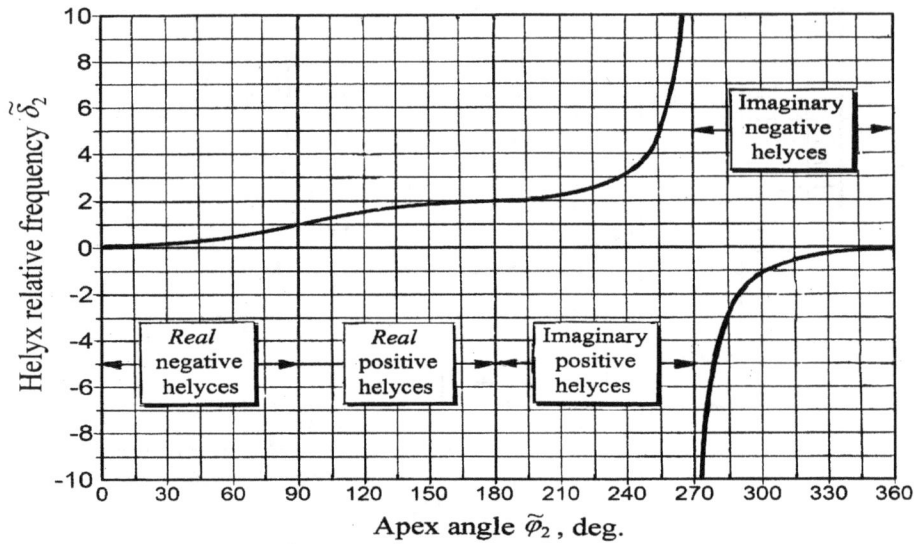

$\widetilde{\varphi}_2$	$360/0^0$	90^0	180^0	270^0
$\widetilde{\delta}_2$	$-0i/+0$	1.0	2.0	$+\infty/-\infty$

Figure 15.6. Relative frequency of trailing strings $\widetilde{\delta}_2$.

15.7 Topological Inversions of Helyx Parameters

The trends of helyx spacetime parameters shown in Figs. 15.2 – 15.6 reveal its unique topological properties. As the spiral length of helyx leading string decreases from positive to negative infinity, the apex angle of helyx trailing string $\widetilde{\varphi}_2$ increases from 0 to 360 degrees.

Consequently, the following four kinds of topological transformations the helyx occur:

- Real negative helyces - ***Wavelength of trailing string*** becomes inverted at $\widetilde{\varphi}_2 = 90^0$
- Real positive helyces - ***Radius of trailing string*** becomes inverted at $\widetilde{\varphi}_2 = 180^0$
- Imaginary positive helyces - ***Spiral length of trailing string*** becomes inverted at $\widetilde{\varphi}_2 = 270^0$
- Imaginary negative helyces - ***Entire helyx*** becomes inverted when $\widetilde{\varphi}_2 = 0^0 / 360^0$.

Figure 15.7 shows the helyx cross-sections, including the radius $\widetilde{b}_2$ and the translation velocity $\widetilde{\beta}_{2t}$ of helyx trailing string, for the apex angle of helyx trailing strings $\widetilde{\varphi}_2$ equal to $0^0/360^0$, 60^0, 90^0, 120^0, 180^0, 240^0, 270^0 and 300^0.

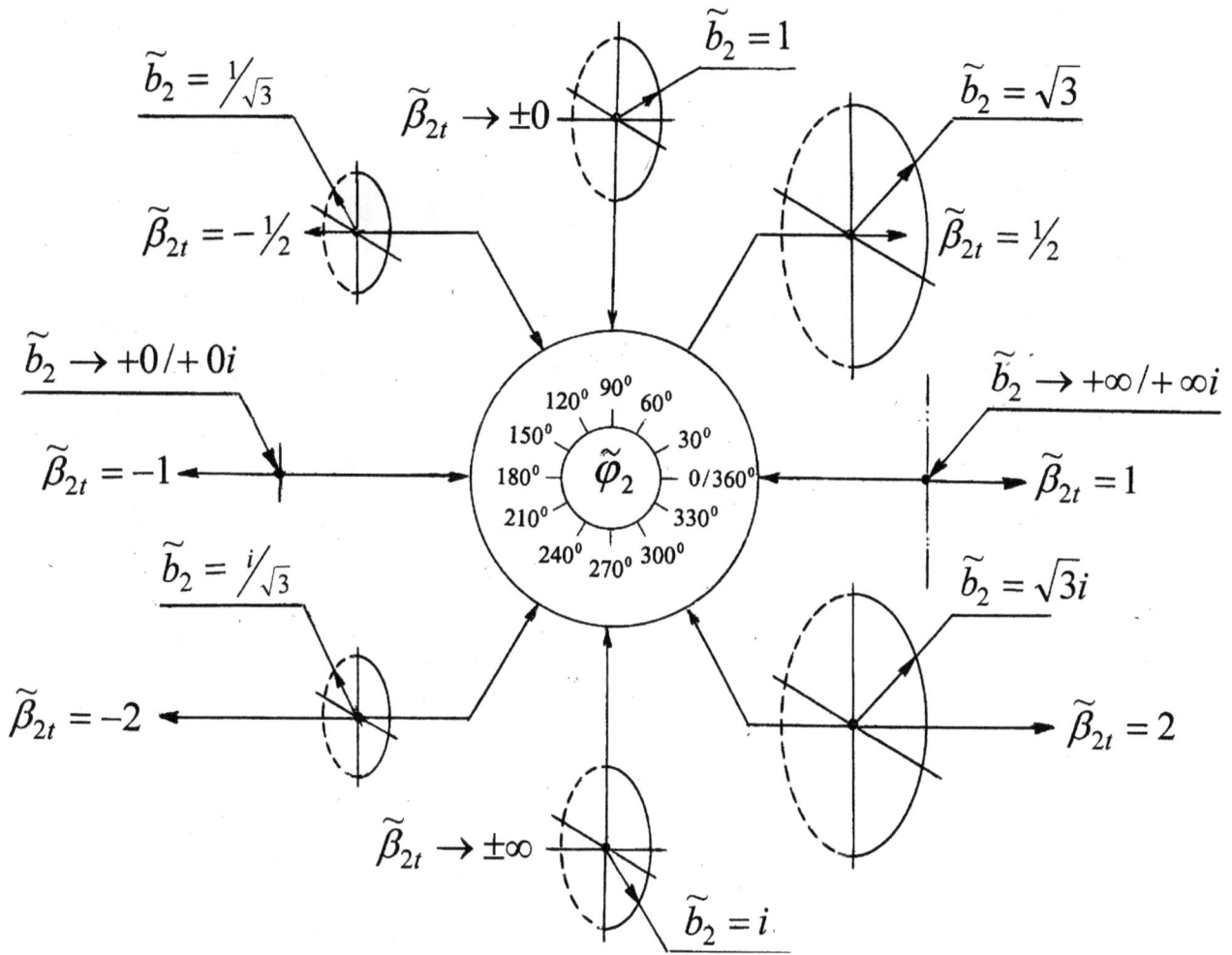

Figure 15.7. Helyx cross-sections and parameters.

<u>Notes:</u>

PART 4

Applied
Mathematics
of a Helyx

16. Basic Elementary Radiation Particles

16.1 Formation of Radiation Particles

Elementary radiation particles are formed when the quantum states m, n and q of their paternal trons and toryces are reduced to the lower levels. In the symbols of emitted radiation particles and their constituent helyces, the right superscripts indicate respectively the quantum states m, n and q of their parental trons and toryces prior to and after their emission of radiation particles. Preceding the right top superscripts are the signs of the radiation particle and helyx vorticities. The absence of any signs indicates that the total vorticity of the radiation particle is equal to zero.

The frequencies of emitted helyces are defined based on the *spacetime intensity conservation law* shown in Exhibit 16.1.

Exhibit 16.1. The spacetime intensity conservation law.

The spacetime intensity $\widetilde{I}$ of an emitted helyx is proportional to the relative frequency of helyx trailing string $\widetilde{\delta}_2$. It is equals to a difference between the spacetime intensities I_k and I_j of its parental toryx in the higher and lower quantum states k and j:

$$\widetilde{I} = 4\alpha_s^{-1}\widetilde{\delta}_2 = I_k - I_j \tag{16.1-1}$$

Consequently we obtain from Eqs. (7.1-12) and (16.1-1) that the relative frequency of helyx trailing string $\widetilde{\delta}_2$ is equal to:

$$\widetilde{\delta}_2 = \frac{Q_{qk}G_k - Q_{qj}G_j}{4\alpha_s^{-1}} \tag{16.1-2}$$

For the helyces emitted by excited parental toryces with the toryx oscillation quantum factors $Q_{qk} = Q_{qj} = 1$. Thus, Eq. (16.1-2) reduces to the form:

$$\widetilde{\delta}_2 = \frac{G_k - G_j}{4\alpha_s^{-1}} \tag{16.1-3}$$

For the helyces emitted by oscillated parental toryces with the toryx golden numbers $G_k = G_j = 1$. Thus, Eq. (16.1-2) reduces to the form:

$$\widetilde{\delta}_2 = \frac{Q_{qk} - Q_{qj}}{4\alpha_s^{-1}} \tag{16.1-4}$$

The energy of an emitted helyx E_{jk} when its parental toryx is transferred from the higher quantum state k to the lower quantum state j is equal to:

$$\widetilde{E}_{jk} = \widetilde{f}_2 h \tag{16.1-5}$$

where h = Planck constant.

The elementary radiation particles emitted by the excited trons are called *tons* as shown in Table 16.1. Similarly to the trons, there are two kinds of tons, the ***charge-polarized tons*** and the ***reality-polarized tons***. Both of them are formed when their respective polarized parental trons are transferred from higher to lower quantum states.

Table 16.1. Types of elementary radiation particles.

Parental matter particles		Elementary radiation particles	
Combined	Constituent	Constituent	Combined
ce-tron	Electron	Electon	ce-ton
	Positron	Positon	
ca-tron	Ethertron	Etherton	ca-ton
	Singulatron	Singulaton	

The formation of two kinds of combined radiation particles are considered below: the combined ce-tons and the combined ca-tons.

The combined ce-tons - The combined ce-tons $\widetilde{ce}_{m,n,q}^{m,n,q}$ are emitted by the combined parental ce-trons $\widetilde{ce}_{m,n,q}^{0}$. The process of creation of the emitted combined ce-ton involves the following steps (see: Sections 8.15, 8.16, 8.19 and Fig. 16.1.1):

- The electron $e_{m,n,q}^{-1}$ of the combined parental ce-tron $\widetilde{ce}_{m,n,q}^{0}$ emits the real and imaginary negative helyces $\widetilde{E}_{m,n,q}^{-m,n,q}$ and $\widetilde{\widetilde{E}}_{m,n,q}^{-m,n,q}$.

- The unification of the real and imaginary negative helyces $\widetilde{E}_{m,n,q}^{-m,n,q}$ and $\widetilde{\widetilde{E}}_{m,n,q}^{-m,n,q}$ produces the *electon* $e_{m,n,q}^{-m,n,q}$.

- The positron $e_{m,n,q}^{+1}$ of the combined parental ce-tron $\widetilde{ce}_{m,n,q}^{0}$ emits the real and imaginary positive helyces $\widetilde{E}_{m,n,q}^{+m,n,q}$ and $\widetilde{\widetilde{E}}_{m,n,q}^{+m,n,q}$.

- The unification of the real and imaginary positive helyces $\widetilde{E}_{m,n,q}^{+m,n,q}$ and $\widetilde{\widetilde{E}}_{m,n,q}^{+m,n,q}$ produces the *positon* $e_{m,n,q}^{+m,n,q}$.

- The combined ce-ton $\widetilde{e}_{m,n,q}^{m,n,q}$ is produced by the unification of the electon $e_{m,n,q}^{-m,n,q}$ with the positon $e_{m,n,q}^{+m,n,q}$ both propagating simultaneously along the same path.

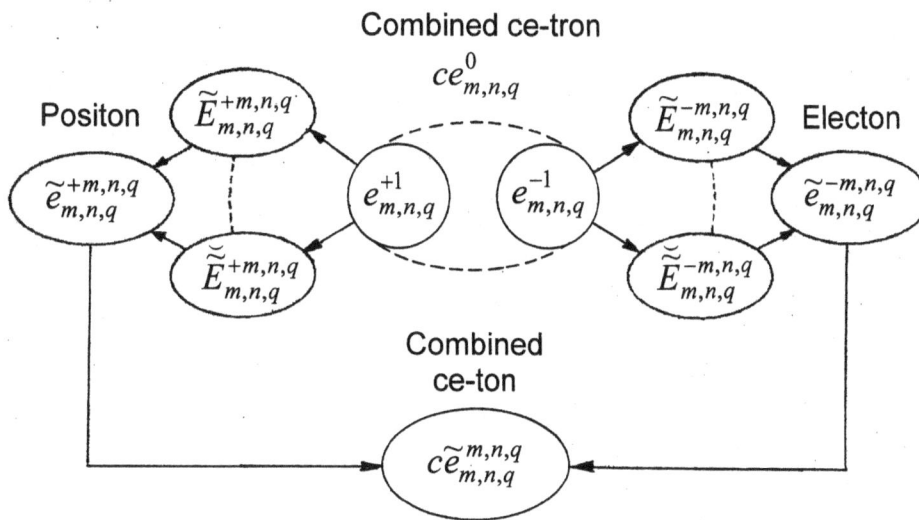

Fig. 16.1.1. Formation of the emitted combined ce-ton $\widetilde{ce}_{m,n,q}^{m,n,q}$ from the parental combined ce-tron $\widetilde{ce}_{m,n,q}^{0}$.

The combined ca-tons – The combined ca-tons $\widetilde{ca}_{m,n,q}^{m,n,q}$ are emitted by the combined parental ca-trons $\widetilde{ca}_{m,n,q}^{0}$. The process of creation of the emitted combined ca-ton involves the following steps (see: Sections 8.17, 8.18. 8.20 and Fig. 16.1.2).

- The singulatron $\breve{\tilde{a}}^0_{m,n,q}$ of the combined ca-tron $c\breve{\tilde{a}}^0_{m,n,q}$ emits the imaginary positive and negative helyces $\breve{\tilde{A}}^{+m,n,q}_{m,n,q}$ and $\breve{\tilde{A}}^{-m,n,q}_{m,n,q}$.

- The unification of the imaginary positive and negative helyces $\breve{\tilde{A}}^{+m,n,q}_{m,n,q}$ and $\breve{\tilde{A}}^{-m,n,q}_{m,n,q}$ produces the ***singulaton*** $\breve{\tilde{a}}^{m,n,q}_{m,n,q}$.

- N excited ethertrons $\tilde{a}^0_{m,n,q}$ of the combined ca-tron $c\tilde{a}^0_{m,n,q}$ emit N positive and negative helyces $\tilde{A}^{+m,n,q}_{m,n,q}$ and $\tilde{A}^{-m,n,q}_{m,n,q}$.

- The unification of N real positive and negative helyces $\tilde{A}^{+m,n,q}_{m,n,q}$ and $\tilde{A}^{-m,n,q}_{m,n,q}$ produces N ***ethertons*** $\tilde{a}^{m,n,q}_{m,n,q}$.

- The combined ca-ton $c\tilde{a}^{m,n,q}_{m,n,q}$ is produced by the unification of one singulaton $\breve{\tilde{a}}^{m,n,q}_{m,n,q}$ with N ethertons $\tilde{a}^{m,n,q}_{m,n,q}$ propagating simultaneously along the same path.

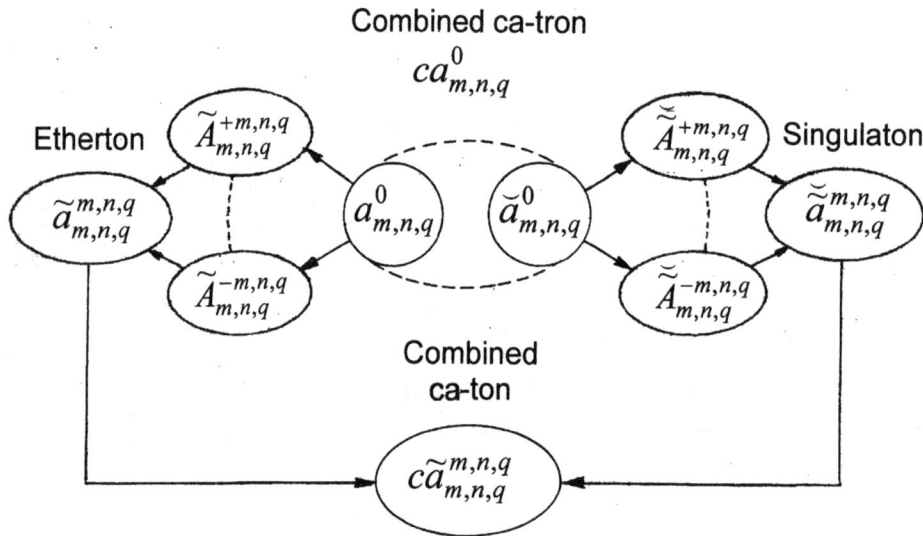

Fig. 16.1.2. Formation of the emitted combined ca-ton $c\tilde{a}^{m,n,q}_{m,n,q}$ from the combined parental ca-tron $c\tilde{a}^0_{m,n,q}$.

16.2 Excited Lambda Electons & Positons

Parental excited lambda electrons - Table 16.2.1 shows compositions and properties of the parental excited lambda electrons $e^{-1}_{m,n,q}$.

Table 16.2.1. Compositions and properties of parental excited lambda electrons of the spacetime levels L with the quantum states (m) when $Q_q = 1$.

L (m)	Parental electrons	Toryces of parental excited lambda electrons					
		Symbols	b_1	$\beta_1 = \beta_{2t}$	β_{2r}	δ_2	G
$L1$ (1)	$e_{1,1,0}^{-1}$	$E_{1,1,0}^{-}$	274.00	0.08535778	0.99635036	3.6496×10^{-3}	-7.3126×10^{-3}
		$\breve{E}_{1,1,0}^{-}$	-273.00	$-0.08567044i$	1.00366300	-3.6630×10^{-3}	7.3126×10^{-3}
	$e_{1,2,0}^{-1}$	$E_{1,2,0}^{-}$	548.00	0.06038464	0.99817518	1.8248×10^{-3}	-3.6530×10^{-3}
		$\breve{E}_{1,2,0}^{-}$	-547.00	$-0.06049504i$	1.00182815	-1.8282×10^{-3}	3.6530×10^{-3}
	$e_{1,3,0}^{-1}$	$E_{1,3,0}^{-}$	822.00	0.04931136	0.99878346	1.2166×10^{-3}	-2.4346×10^{-3}
		$\breve{E}_{1,3,0}^{-}$	-821.00	$-0.04937142i$	1.00181218	-1.2180×10^{-3}	2.4346×10^{-3}
$L2$ (2)	$e_{2,1,0}^{-1}$	$E_{2,1,0}^{-}$	37538.0	0.00729922	0.99997336	2.6640×10^{-5}	-5.3280×10^{-5}
		$\breve{E}_{2,1,0}^{-}$	-37537.0	$-0.00729942i$	1.00002664	-2.6640×10^{-5}	5.3280×10^{-5}
	$e_{2,2,0}^{-1}$	$E_{2,2,0}^{-}$	150152.0	0.00364963	0.99999334	6.6600×10^{-6}	-1.3320×10^{-5}
		$\breve{E}_{2,2,0}^{-}$	-150151.0	$-0.00364965i$	1.00000666	-6.6600×10^{-6}	1.3320×10^{-5}
	$e_{2,3,0}^{-1}$	$E_{2,3,0}^{-}$	337842.0	0.00243309	0.99999970	2.9600×10^{-6}	-5.9120×10^{-6}
		$\breve{E}_{2,3,0}^{-}$	-337841.0	$-0.00243310i$	1.00000296	-2.9600×10^{-6}	5.9120×10^{-6}
$L3$ (3)	$e_{3,1,0}^{-1}$	$E_{3,1,0}^{-}$	5142706.0	0.00062362	0.99999981	1.9445×10^{-7}	-1.9445×10^{-7}
		$\breve{E}_{3,1,0}^{-}$	-5142705.0	$-0.00062362i$	1.00000019	-1.9445×10^{-7}	1.9445×10^{-7}
	$e_{3,2,0}^{-1}$	$E_{3,2,0}^{-}$	41141648.0	0.00022048	0.99999998	2.4306×10^{-8}	-4.8612×10^{-8}
		$\breve{E}_{3,2,0}^{-}$	-41141647.0	$-0.00022048i$	1.00000002	-2.4306×10^{-8}	4.8612×10^{-8}
	$e_{3,3,0}^{-1}$	$E_{3,3,0}^{-}$	138853062.0	0.00012002	0.99999993	7.2018×10^{-9}	-1.4404×10^{-8}
		$\breve{E}_{3,3,0}^{-}$	-138853061.0	$-0.00012002i$	1.00000007	-7.2018×10^{-9}	1.4404×10^{-8}

Emitted excited lambda electons - Table 16.2.2 shows compositions and properties of the emitted excited lambda electons $\widetilde{e}_{m,n,q}^{-m,n,q}$.

Table 16.2.2. Compositions and properties of emitted excited lambda electrons of the spacetime levels L with the quantum states (m).

L (m)	Emitted electrons	Symbols	$\tilde{b}_2$	$\tilde{\beta}_{2t}$	$\tilde{\beta}_{2r}$	$\tilde{f}_2$, Hz	$\tilde{E}_2$, MeV
L1 (1)	$\tilde{e}_{1,1,0}^{-1,2,0}$	$\tilde{E}_{1,1,0}^{-1,2,0}$	5.473×10^2	0.99999332	0.00365416	2.26×10^{17}	9.35×10^{-4}
		$\widetilde{\widetilde{E}}_{1,1,0}^{-1,2,0}$	$5.473 \times 10^2 i$	1.00000668	$-0.00365417i$	-2.26×10^{17}	-9.35×10^{-4}
	$\tilde{e}_{1,2,0}^{-1,3,0}$	$\tilde{E}_{1,2,0}^{-1,3,0}$	9.486×10^2	0.99999778	0.00210844	7.53×10^{16}	3.11×10^{-4}
		$\widetilde{\widetilde{E}}_{1,2,0}^{-1,3,0}$	$9.486 \times 10^2 i$	1.00000222	$-0.00210845i$	-7.53×10^{16}	-3.11×10^{-4}
L2 (2)	$\tilde{e}_{2,1,0}^{-2,2,0}$	$\tilde{E}_{2,1,0}^{-2,2,0}$	5.238×10^3	0.999999927	0.00038184	2.47×10^{15}	1.02×10^{-5}
		$\widetilde{\widetilde{E}}_{2,1,0}^{-2,2,0}$	$5.238 \times 10^4 i$	1.000000073	$-0.00038184i$	-2.47×10^{15}	-1.02×10^{-5}
	$\tilde{e}_{2,2,0}^{-2,3,0}$	$\tilde{E}_{2,2,0}^{-2,3,0}$	1.217×10^4	0.999999987	0.00016432	4.57×10^{14}	1.89×10^{-6}
		$\widetilde{\widetilde{E}}_{2,2,0}^{-2,3,0}$	$1.217 \times 10^4 i$	1.000000013	$-0.00016432i$	-4.57×10^{14}	-1.89×10^{-6}
L3 (3)	$\tilde{e}_{2,2,0}^{-3,2,0}$	$\tilde{E}_{3,1,0}^{-3,2,0}$	5.676×10^4	0.9999999994	0.00003524	2.10×10^{13}	8.69×10^{-8}
		$\widetilde{\widetilde{E}}_{3,1,0}^{-3,2,0}$	$5.676 \times 10^4 i$	1.0000000006	$-0.00003524i$	-2.10×10^{13}	-8.69×10^{-8}
	$\tilde{e}_{3,2,0}^{-3,3,0}$	$\tilde{E}_{3,2,0}^{-3,3,0}$	1.790×10^5	0.9999999999	0.00001117	2.11×10^{12}	8.74×10^{-9}
		$\widetilde{\widetilde{E}}_{3,2,0}^{-3,3,0}$	$1.790 \times 10^5 i$	1.0000000001	$-0.00001117i$	-2.11×10^{12}	-8.74×10^{-9}

Parental excited lambda positrons - Table 16.2.3 shows compositions and properties of the parental excited lambda positrons $e_{m,n,q}^{+1}$.

Table 16.2.3. Compositions and properties of parental excited lambda positrons of the spacetime levels L with the quantum states (m) when $Q_q = 1$.

L (m)	Parental positrons	\multicolumn{6}{c}{Toryces of parental excited lambda positrons}
		Symbols	b_1	$\beta_1 = \beta_{2t}$	β_{2r}	δ_2	G
$L1$ (1)	$e^{+1}_{1,1,0}$	$E^+_{1,1,0}$	0.50091408	0.08535778	-0.99635036	1.99635036	7.3126×10^{-3}
		$\breve{E}^+_{1,1,0}$	0.49908592	$0.08567044i$	-1.00366300	2.00366300	-7.3126×10^{-3}
	$e^{+1}_{1,2,0}$	$E^+_{1,2,0}$	0.50045662	0.06038464	-0.99817518	1.99817518	3.6530×10^{-3}
		$\breve{E}^+_{1,2,0}$	0.49954338	$0.06049504i$	-1.00182815	2.00182815	-3.6530×10^{-3}
	$e^{+1}_{1,3,0}$	$E^+_{1,3,0}$	0.50030432	0.04931136	-0.99878346	1.99878345	2.4346×10^{-3}
		$\breve{E}^+_{1,3,0}$	0.49969568	$0.04937142i$	-1.00181218	2.00121803	-2.4346×10^{-3}
$L2$ (2)	$e^{+1}_{2,1,0}$	$E^+_{2,1,0}$	0.50000666	0.00729922	-0.99997336	1.99997336	5.3280×10^{-5}
		$\breve{E}^+_{2,1,0}$	0.49999334	$0.00729942i$	-1.00002664	2.00002664	-5.3280×10^{-5}
	$e^{+1}_{2,2,0}$	$E^+_{2,2,0}$	0.50000166	0.00364963	-0.99999334	1.99999334	1.3320×10^{-5}
		$\breve{E}^+_{2,2,0}$	0.49999834	$0.00364965i$	-1.00000666	2.00000666	-1.3320×10^{-5}
	$e^{+1}_{2,3,0}$	$E^+_{2,3,0}$	0.50000074	0.00243309	-0.99999704	1.99999704	5.9120×10^{-6}
		$\breve{E}^+_{2,3,0}$	0.49999926	$0.00243310i$	-1.00000296	2.00000296	-5.9120×10^{-6}
$L3$ (3)	$e^{+1}_{3,1,0}$	$E^+_{3,1,0}$	0.5000000486	0.00062362	-0.99999981	1.999999805	1.9445×10^{-7}
		$\breve{E}^+_{3,1,0}$	0.4999999514	$0.00062362i$	-1.00000019	2.000000195	-1.9445×10^{-7}
	$e^{+1}_{3,2,0}$	$E^+_{3,2,0}$	0.5000000061	0.00022048	-0.99999998	1.999999976	4.8612×10^{-8}
		$\breve{E}^+_{3,2,0}$	0.4999999939	$0.00022048i$	-1.00000002	2.000000024	-4.8612×10^{-8}
	$e^{+1}_{3,3,0}$	$E^+_{3,3,0}$	0.5000000018	0.00012002	-0.99999993	1.999999993	1.4404×10^{-8}
		$\breve{E}^+_{3,3,0}$	0.4999999982	$0.00012002i$	-1.00000007	2.000000007	-1.4404×10^{-8}

Emitted excited lambda positons - Table 16.2.4 shows compositions and properties of the excited lambda positons $\widetilde{e}^{+m,n,q}_{m,n,q}$.

Table 16.2.4. Compositions and properties of emitted excited lambda positons of the spacetime levels L with the quantum states (m).

L (m)	Emitted positons	\multicolumn{6}{c}{Helyces of emitted excited lambda positons}
		Symbols	$\tilde{b}_2$	$\tilde{\beta}_{2t}$	$\tilde{\beta}_{2r}$	$\tilde{f}_2$, Hz	$\tilde{E}_2$, MeV
L1 (1)	$\tilde{e}_{1,1,0}^{+1,2,0}$	$\tilde{\tilde{E}}_{1,1,0}^{+1,2,0}$	$5.473\times10^2 i$	1.00000668	$-0.00365417i$	-2.26×10^{17}	-9.35×10^{-4}
		$\tilde{E}_{1,1,0}^{+1,2,0}$	5.473×10^2	0.99999332	0.00365416	2.26×10^{17}	9.35×10^{-4}
	$\tilde{e}_{1,2,0}^{+1,3,0}$	$\tilde{\tilde{E}}_{1,2,0}^{+1,3,0}$	$9.486\times10^2 i$	1.00000222	$-0.00210845i$	-7.53×10^{16}	-3.11×10^{-4}
		$\tilde{E}_{1,2,0}^{+1,3,0}$	9.486×10^2	0.99999778	0.00210844	7.53×10^{16}	3.11×10^{-4}
L2 (2)	$\tilde{e}_{2,1,0}^{+2,2,0}$	$\tilde{\tilde{E}}_{2,1,0}^{+2,2,0}$	$5.238\times10^4 i$	1.000000073	$-0.00038184i$	-2.47×10^{15}	-1.02×10^{-5}
		$\tilde{E}_{2,1,0}^{+2,2,0}$	5.238×10^3	0.999999927	0.00038184	2.47×10^{15}	1.02×10^{-5}
	$\tilde{e}_{2,2,0}^{+2,3,0}$	$\tilde{\tilde{E}}_{2,2,0}^{+2,3,0}$	$1.217\times10^4 i$	1.000000013	$-0.00016432i$	-4.57×10^{14}	-1.89×10^{-6}
		$\tilde{E}_{2,2,0}^{+2,3,0}$	1.217×10^4	0.999999987	0.00016432	4.57×10^{14}	1.89×10^{-6}
L3 (3)	$\tilde{e}_{2,2,0}^{+3,2,0}$	$\tilde{\tilde{E}}_{3,1,0}^{+3,2,0}$	$5.676\times10^4 i$	1.0000000006	$-0.00003524i$	-2.10×10^{13}	-8.69×10^{-8}
		$\tilde{E}_{3,1,0}^{+3,2,0}$	5.676×10^4	0.9999999994	0.00003524	2.10×10^{13}	8.69×10^{-8}
	$\tilde{e}_{3,2,0}^{+3,3,0}$	$\tilde{\tilde{E}}_{3,2,0}^{+3,3,0}$	$1.790\times10^5 i$	1.0000000001	$-0.00001117i$	-2.11×10^{12}	-8.74×10^{-9}
		$\tilde{E}_{3,2,0}^{+3,3,0}$	1.790×10^5	0.9999999999	0.00001117	2.11×10^{12}	8.74×10^{-9}

Analysis of calculated data for excited lambda electons & positons - It follows from Tables 16.2.2 - 16.2.4 and Fig. 16.1.1:

- Real and imaginary helyces making up both excited electons and positons are emitted in the same direction, but they have opposite spins.
- The translational velocities of trailing strings $\tilde{\beta}_{2t}$ of emitted real helyces making up excited electons and positons are slightly less than velocity of light, while these velocities of emitted imaginary helyces are slightly greater than velocity of light.

The following changes of parameters of trailing strings of constituent helyces of excited lambda electons and positons occur as the spacetime level L increases:

- The relative radius $\widetilde{b}_2$ increases
- The frequency $\widetilde{f}_2$ decreases
- The differences between the translational velocities of trailing strings $\widetilde{\beta}_{2t}$ of real and imaginary helyces decrease.

The calculated frequencies of electons and positons $\widetilde{f}_2$ of various spacetime levels L are within the following frequency ranges:

- For the spacetime level *L1* ($m = 1$), the calculated frequencies are within the frequency range of cosmic **X-ray background (CXB) radiations**.
- For the spacetime level *L2* ($m = 2$), the calculated frequencies are within the frequency ranges of **infrared, visible** and **ultraviolet radiations**.
- For the spacetime level *L3* ($m = 3$), the calculated frequencies are within the frequency ranges of infrared and microwave frequency ranges known in astronomy as **cosmic microwave background (CMB) radiation**.

Experimental data for the frequencies $\widetilde{f}_2$ of some spectra lines of a hydrogen atom are very accurately described by the Rydberg's equation:

$$\widetilde{f}_2 = R_\infty c \left(\frac{1}{n_j^2} - \frac{1}{n_k^2} \right)$$

(16.2-1)

where R_∞ is the Rydberg constant.

Table 16.2.5. Comparison of the calculated frequencies $\widetilde{f}_2$ of emitted electon-positons with frequencies of spectra lines for hydrogen atom calculated from the Rydberg's equation.

Spectra lines of hydrogen	Quantum states ($k-j$)	Excited ce-tons	Frequencies $\widetilde{f}_2$, Hz		Calculated/ Rydbers's
			Calculated	Rydberg's	
H_α	$(3-2)$	$c\widetilde{e}_{2,2,0}^{2,3,0}$	4.571648×10^{14}	4.569225×10^{14}	1.000530
H_β	$(4-2)$	$c\widetilde{e}_{2,2,0}^{2,4,0}$	6.171721×10^{14}	6.168454×10^{14}	1.000530
H_γ	$(5-2)$	$c\widetilde{e}_{2,2,0}^{2,5,0}$	6.912325×10^{14}	6.908668×10^{14}	1.000529
H_δ	$(6-2)$	$c\widetilde{e}_{2,2,0}^{2,6,0}$	7.314629×10^{14}	7.310760×10^{14}	1.000529

Table 16.2.5 shows a comparison of the calculated frequencies $\tilde{f}_2$ of emitted electon-positons with frequencies of spectra lines for hydrogen atom calculated from the Rydberg's equation.

16.3 Excited Harmonic Electons & Positons

Parental excited harmonic electrons - Table 16.3.1 shows compositions and properties of the parental harmonic electrons $e_{H,n,q}^{-1}$ of all spacetime levels L.

Table 16.3.1. Compositions and properties of parental excited harmonic electrons of all spacetime levels L with the quantum states $m = 0$.

Parental electrons	Toryces of parental excited harmonic electrons					
	Symbols	b_1	$\beta_1 = \beta_{2t}$	β_{2r}	δ_2	G
$e_{H,0,0}^{-1}$	$E_{H,0,0}^{-1/2}$	2.00	0.86602540	0.50000000	0.50000000	-1.50000000
	$\breve{E}_{H,0,0}^{-2}$	-1.00	-1.73205081i	2.00000000	-1.00000000	1.50000000
$e_{H,1,0}^{-1}$	$E_{H,1,0}^{-2/3}$	3.00	0.74535599	0.66666667	0.33333333	-0.83333333
	$\breve{E}_{H,1,0}^{-3/2}$	-2.00	-1.11803399i	1.50000000	-0.50000000	0.83333333
$e_{H,2,0}^{-1}$	$E_{H,2,0}^{-3/4}$	4.00	0.66143783	0.75000000	0.25000000	-0.58333333
	$\breve{E}_{H,2,0}^{-4/3}$	-3.00	-0.88191710i	1.33333333	-0.33333333	0.58333333
$e_{H,3,0}^{-1}$	$E_{H,3,0}^{-4/5}$	5.0	0.60000000	0.80000000	0.20000000	-0.45000000
	$\breve{E}_{H,3,0}^{-5/4}$	-4.0	-0.75000000i	1.25000000	-0.25000000	0.45000000

Emitted excited harmonic electons - Tables 16.3.2 shows compositions and properties of the emitted harmonic electons $\tilde{e}_{H,n,q}^{-H,n,q}$ of all spacetime levels L.

Table 16.3.2. Compositions and properties of emitted excited harmonic electons of all spacetime levels L with the quantum states $m = 0$.

Emitted electons	Symbols	Helyces of emitted excited harmonic electons				
		$\tilde{b}_2$	$\tilde{\beta}_{2t}$	$\tilde{\beta}_{2r}$	$\tilde{f}_2$, Hz	$\tilde{E}_2$, MeV
$e^{-H,1,0}_{H,0,0}$	$\tilde{E}^{-H,1,0}_{H,0,0}$	28.656868	0.99756755	0.06970642	8.237×10^{19}	0.340666
	$\breve{\tilde{E}}^{-1,1,0}_{1,0,0}$	28.691743i	1.00243245	-0.6979125i	-8.237×10^{19}	-0.340666
$e^{-H,2,0}_{H,1,0}$	$\tilde{E}^{-H,2,0}_{H,1,0}$	46.814274	0.99908783	0.04270253	3.089×10^{19}	0.127750
	$\breve{\tilde{E}}^{-H,2,0}_{H,1,0}$	46.835630i	1.00091217	-0.04272201i	-3.089×10^{19}	-0.340666
$e^{-H,3,0}_{H,2,0}$	$\tilde{E}^{-H,3,0}_{H,2,0}$	64.109909	0.99951351	0.03118884	1.648×10^{19}	0.068133
	$\breve{\tilde{E}}^{-H,3,0}_{H,2,0}$	64.125506i	1.00048649	-0.03119642i	-1.648×10^{19}	-0.340666

Parental excited harmonic positrons - Table 16.3.3 shows compositions and properties of the parental excited harmonic positrons.

Table 16.3.3. Compositions and properties of parental excited harmonic positrons of all spacetime levels L with the quantum states $m = 0$.

Parental positrons	Symbols	Toryces of parental excited harmonic positrons				
		b_1	$\beta_1 = \beta_{2t}$	β_{2r}	δ_2	G
$e^{+1}_{H,0,0}$	$E^{+1/2}_{H,0,0}$	0.66666667	0.86602540	-0.50000000	1.50000000	1.50000000
	$\breve{E}^{+2}_{H,0,0}$	0.33333333	1.73205081i	-2.00000000	3.00000000	-1.50000000
$e^{+1}_{H,1,0}$	$E^{+2/3}_{H,1,0}$	0.60000000	0.74535599	-0.66666667	1.66666667	0.83333333
	$\breve{E}^{+3/2}_{H,1,0}$	0.40000000	1.11803399i	-1.50000000	2.50000000	-0.83333333
$e^{+1}_{H,2,0}$	$E^{+3/4}_{H,2,0}$	0.57142857	0.66143783	-0.75000000	1.75000000	0.58333333
	$\breve{E}^{+4/3}_{H,2,0}$	0.42857143	0.88191710i	-1.33333333	2.33333333	-0.58333333
$e^{+1}_{H,3,0}$	$E^{+4/5}_{H,3,0}$	0.55555556	0.60000000	-0.80000000	1.80000000	0.45000000
	$\breve{E}^{+5/4}_{H,3,0}$	0.44444444	0.75000000i	-1.25000000	2.25000000	-0.45000000

Emitted excited harmonic positons - Table 16.3.4 shows compositions and properties of the excited harmonic positons $\widetilde{e}^{+H,n,q}_{H,n,q}$ of all spacetime levels L.

Table 16.3.4. Compositions and properties of emitted excited harmonic positons of all spacetime levels L with the quantum states $m = 0$.

Emitted positons	\ Symbols	\ $\widetilde{b}_2$	\ $\widetilde{\beta}_{2t}$	\ $\widetilde{\beta}_{2r}$	\ $\widetilde{f}_2$, Hz	\ $\widetilde{E}_2$, MeV
$e^{+H,1,0}_{H,0,0}$	$\widetilde{E}^{+H,1,0}_{H,0,0}$	$28.691743i$	1.00243245	$-0.6979125i$	-8.237×10^{19}	-0.340666
	$\widetilde{E}^{+1,1,0}_{1,0,0}$	28.656868	0.99756755	0.06970642	8.237×10^{19}	0.340666
$e^{+H,2,0}_{H,1,0}$	$\widetilde{E}^{+H,2,0}_{H,1,0}$	$46.835630i$	1.00091217	$-0.04272201i$	-3.089×10^{19}	-0.340666
	$\widetilde{E}^{+H,2,0}_{H,1,0}$	46.814274	0.99908783	0.04270253	3.089×10^{19}	0.127750
$e^{+H,3,0}_{H,2,0}$	$\widetilde{E}^{+H,3,0}_{H,2,0}$	$64.125506i$	1.00048649	$-0.03119642i$	-1.648×10^{19}	-0.340666
	$\widetilde{E}^{+H,3,0}_{H,2,0}$	64.109909	0.99951351	0.03118884	1.648×10^{19}	0.068133

Analysis of calculated data for excited harmonic electons & positons - It follows from Tables 16.3.2 - 16.3.4 and Fig. 16.1.1:

- Real and imaginary helyces making up both excited electons and positons are emitted in the same direction, but they have opposite spins.
- The translational velocities of trailing strings $\widetilde{\beta}_{2t}$ of emitted real helyces making up excited electons and positrons are slightly less than velocity of light, while these velocities of emitted imaginary helyces are slightly greater than velocity of light.

The calculated frequencies of electons and positons $\widetilde{f}_2$ are independent on the spacetime levels L and they are within the frequency range of the **X-ray electromagnetic waves**.

16.4 Excited Lambda Singulatons & Ethertons

Parental singulatrons of excited lambda singulatons - Table 16.4.1 shows compositions and properties of the parental excited lambda singulatrons $\widetilde{a}^{0}_{m,n,q}$.

Table 16.4.1. Compositions and properties of parental excited lambda singulatrons of the spacetime levels L with the quantum states (m).

L (m)	Parental singulatrons	Symbols	b_1	$\beta_1 = \beta_{2t}$	β_{2r}	δ_2	G
L1 (0)	$\breve{a}^0_{H,0,0}$	$\breve{A}^-_{H,0,0}$	-1.00000000	-1.73205081i	2.00000000	-1.00000000	1.50000000
		$\breve{A}^+_{H,0,0}$	0.33333333	1.73205081i	-2.00000000	3.00000000	-1.50000000
	$\breve{a}^0_{H,0,0}$	$\breve{A}^-_{H,1,0}$	-0.50000000	-2.82842712i	3.00000000	-2.00000000	2.66666667
		$\breve{A}^+_{H,1,0}$	0.25000000	2.82842712i	-3.00000000	4.00000000	-2.66666667
	$\breve{a}^0_{H,1,0}$	$\breve{A}^-_{H,2,0}$	-0.33333333	-3.87298335i	4.00000000	-3.00000000	3.75000000
		$\breve{A}^+_{H,2,0}$	0.20000000	3.87298335i	-4.00000000	5.00000000	-3.75000000
L2 (1)	$\breve{a}^0_{1,1,0}$	$\breve{A}^-_{1,1,0}$	-0.00366300	-273.998175i	274.000000	-273.000000	273.996350
		$\breve{A}^+_{1,1,0}$	0.00363636	273.998175i	-274.000000	275.000000	-273.996350
	$\breve{a}^0_{1,2,0}$	$\breve{A}^-_{1,2,0}$	-0.00182815	-547.999088i	548.000000	-547.000000	574.998175
		$\breve{A}^+_{1,2,0}$	0.00182149	547.999088i	-548.000000	549.000000	-574.998175
	$\breve{a}^0_{1,3,0}$	$\breve{A}^-_{1,3,0}$	-0.00121803	-821.999392i	822.000000	-821.000000	821.998784
		$\breve{A}^+_{1,3,0}$	0.00121507	821.999392i	-822.000000	823.000000	-821.998784
L3 (2)	$\breve{a}^0_{2,1,0}$	$\breve{A}^-_{2,1,0}$	-0.00002664	-37538.0000i	37538.0000	-37537.0000	37538.000
		$\breve{A}^+_{2,1,0}$	0.00002664	37538.0000i	-37538.0000	37539.0000	-37538.000
	$\breve{a}^0_{2,2,0}$	$\breve{A}^-_{2,2,0}$	-0.00000666	-152152.000i	152152.000	-152151.000	152152.000
		$\breve{A}^+_{2,2,0}$	0.00000666	152152.000i	-152152.000	152153.000	-152152.000
	$\breve{a}^0_{2,3,0}$	$\breve{A}^-_{2,3,0}$	-0.00000296	-337842.000i	337842.000	-337841.000	337842.000
		$\breve{A}^+_{2,3,0}$	0.00000296	337842.000i	-337842.000	337843.000	-337842.000

Emitted excited lambda singulatons - Table 16.4.2 and Fig. 16.1.2 show compositions, properties and formation of the emitted excited lambda singulatons $\widetilde{a}^{m,n,q}_{m,n,q}$.

Table 16.4.2. Compositions and properties of emitted excited lambda singulatons of the spacetime levels L with the quantum states (m).

L (m)	Emitted singula-tons	Symbols	$\tilde{b}_2$	$\tilde{\beta}_{2t}$	$\tilde{\beta}_{2r}$	$\tilde{f}_2$, Hz	$\tilde{E}_2$, GeV
L1 (0)	$\tilde{\tilde{a}}^{H,1,0}_{H,0,0}$	$\tilde{\tilde{A}}^{-H,1,0}_{H,0,0}$	21.652662	0.99574321	0.09217080	1.442×10^{20}	0.0005962
		$\tilde{\tilde{A}}^{+H,1,0}_{H,0,0}$	21.698796i	1.00425679	-0.09236719i	$-1.442\times10^{20}i$	-0.0005962
	$\tilde{\tilde{a}}^{H,2,0}_{H,1,0}$	$\tilde{\tilde{A}}^{-H,2,0}_{H,1,0}$	22.471741	0.99604727	0.08882477	1.339×10^{20}	0.0005536
		$\tilde{\tilde{A}}^{+H,2,0}_{H,1,0}$	22.516197i	1.00395273	-0.08900050i	$-1.339\times10^{20}i$	-0.0005536
	$\tilde{\tilde{a}}^{H,3,0}_{H,2,0}$	$\tilde{\tilde{A}}^{-H,3,0}_{H,2,0}$	22.826344	0.99616889	0.08745023	1.297×10^{20}	0.0005366
		$\tilde{\tilde{A}}^{+H,3,0}_{H,2,0}$	22.870111i	1.00383111	-0.08761790i	$-1.297\times10^{20}i$	-0.0005366
L2 (1)	$\tilde{\tilde{a}}^{1,2,0}_{1,1,0}$	$\tilde{\tilde{A}}^{-1,2,0}_{1,1,0}$	1.73234665	0.50012808	0.86559144	1.693×10^{22}	0.0700073
		$\tilde{\tilde{A}}^{+1,2,0}_{1,1,0}$	2.23629714i	1.49987192	-1.11786214i	$-1.693\times10^{22}i$	-0.0700073
	$\tilde{\tilde{a}}^{1,3,0}_{1,2,0}$	$\tilde{\tilde{A}}^{-1,3,0}_{1,2,0}$	1.73235177	0.50013030	0.86595016	1.693×10^{22}	0.0700070
		$\tilde{\tilde{A}}^{+1,3,0}_{1,2,0}$	2.23630111i	1.49986970	-1.11785917i	$-1.693\times10^{22}i$	-0.0700070
	$\tilde{\tilde{a}}^{1,4,0}_{1,3,0}$	$\tilde{\tilde{A}}^{-1,4,0}_{1,3,0}$	1.73235306	0.50013085	0.86594984	1.693×10^{22}	0.0700069
		$\tilde{\tilde{A}}^{+1,4,0}_{1,3,0}$	2.23630211i	1.49986915	-1.11785842i	$-1.693\times10^{22}i$	-0.0700069
L3 (2)	$\tilde{\tilde{a}}^{2,2,0}_{2,1,0}$	$\tilde{\tilde{A}}^{-2,2,0}_{2,1,0}$	0.99512064	-204.445991	204.443545i	6.957×10^{24}	28.772802
		$\tilde{\tilde{A}}^{+2,2,0}_{2,1,0}$	1.00484567i	206.445991	-206.443569i	-6.957×10^{24}	-28.772802
	$\tilde{\tilde{a}}^{2,3,0}_{2,2,0}$	$\tilde{\tilde{E}}^{-2,3,0}_{2,2,0}$	0.99707525	-341.409984	341.408520i	1.160×10^{25}	47.954670
		$\tilde{\tilde{E}}^{+2,3,0}_{2,2,0}$	1.00291622	343.409984	-343.408528i	-1.160×10^{25}	-47.954670
	$\tilde{\tilde{a}}^{2,4,0}_{2,3,0}$	$\tilde{\tilde{A}}^{-2,4,0}_{2,3,0}$	0.99791177	-478.373978	478.372933i	1.623×10^{25}	67.136538
		$\tilde{\tilde{A}}^{+2,4,0}_{2,3,0}$	1.00208388	480.373978	-480.372937i	-1.160×10^{25}	-67.136538

Parental ethertrons of excited lambda ethertons - Table 16.4.3 shows compositions and properties of the parental excited lambda ethertrons $\tilde{\tilde{a}}^0_{m,n,q}$.

Table 16.4.3. Compositions and properties of parental excited lambda ethertrons of the spacetime levels L with the quantum states (m) when $Q_q = 1$.

L (m)	Parental ether-trons	Symbols	b_1	$\beta_1 = \beta_{2t}$	β_{2r}	δ_2	G
$L1$ (0)	$a^0_{H,0,0}$	$A^-_{H,0,0}$	2.00000000	0.86602540	0.50000000	0.50000000	-1.50000000
		$A^+_{H,0,0}$	0.66666667	0.86602540	-0.50000000	1.50000000	1.50000000
	$a^0_{H,1,0}$	$A^-_{H,1,0}$	1.50000000	0.94280904	0.33333333	0.66666667	-2.66666667
		$A^+_{H,1,0}$	0.75000000	0.94280904	-0.33333333	1.33333333	2.66666667
	$a^0_{H,2,0}$	$A^-_{H,2,0}$	1.33333333	0.98624584	0.25000000	0.75000000	-3.75000000
		$A^+_{H,2,0}$	0.80000000	0.98624584	-0.25000000	1.25000000	3.75000000
$L2$ (1)	$a^0_{1,1,0}$	$A^+_{1,1,0}$	1.00363636	0.99999334	3.6496×10^{-3}	0.99635036	-273.996350
		$A^-_{1,1,0}$	0.99636364	0.99999334	-3.6496×10^{-3}	1.00364964	273.996350
	$a^0_{1,2,0}$	$A^+_{1,2,0}$	1.00182149	0.99999834	1.8248×10^{-3}	0.99817518	-574.998175
		$A^-_{1,2,0}$	0.99817851	0.99999834	-1.8248×10^{-3}	1.00182482	574.998175
	$a^0_{1,3,0}$	$A^+_{1,3,0}$	1.00121507	0.99999926	1.2166×10^{-3}	0.99878345	-821.998784
		$A^-_{1,3,0}$	0.99878493	0.99999926	-1.2166×10^{-3}	1.00121655	821.998784
$L3$ (2)	$a^0_{2,1,0}$	$A^+_{2,1,0}$	1.00002664	0.9999999997	2.6640×10^{-5}	0.99635036	-37538.000
		$A^-_{2,1,0}$	0.99997336	0.9999999997	-2.6640×10^{-5}	1.00364964	37538.000
	$a^0_{2,2,0}$	$A^+_{2,2,0}$	1.00000666	0.9999999998	6.6599×10^{-6}	0.99817518	-152152.000
		$A^-_{2,2,0}$	0.99999334	0.9999999998	-6.6599×10^{-6}	1.00182482	152152.000
	$a^0_{2,3,0}$	$A^+_{2,3,0}$	1.00000296	0.9999999999	2.9600×10^{-6}	0.99878345	-337842.000
		$A^-_{2,3,0}$	0.99999704	0.9999999999	-2.9600×10^{-6}	1.00121655	337842.000

Emitted excited lambda ethertons - Table 16.4.4 shows compositions and properties of the emitted excited lambda ethertons $N\widetilde{a}^{m,n,q}_{m,n,q}$.

Table 16.4.4. Compositions and properties of emitted excited lambda ethertons of the spacetime levels L with the quantum states (m).

L (m)	Emitted ethertons	Helyces of emitted excited lambda etherton					
		Symbols	$\tilde{b}_2$	$\tilde{\beta}_{2t}$	$\tilde{\beta}_{2r}$	$\tilde{f}_2$, Hz	$\tilde{E}_2$, GeV
L1 (0)	$\tilde{a}_{H,0,0}^{H,1,0}$	$N\tilde{A}_{H,0,0}^{+H,1,0}$	$21.698796i$	1.00425679	$-0.09236719i$	$-1.442\times10^{20}i$	-0.0005962
		$N\tilde{A}_{H,0,0}^{-H,1,0}$	21.652662	0.99574321	0.09217080	1.442×10^{20}	0.0005962
	$\tilde{a}_{H,1,0}^{H,2,0}$	$N\tilde{A}_{H,1,0}^{+H,2,0}$	$22.516197i$	1.00395273	$-0.08900050i$	$-1.339\times10^{20}i$	-0.0005536
		$\tilde{A}V_{H,1,0}^{-H,2,0}$	22.471741	0.99604727	0.08882477	1.339×10^{20}	0.0005536
	$\tilde{a}_{H,2,0}^{H,3,0}$	$N\tilde{A}_{H,2,0}^{+H,3,0}$	$22.870111i$	1.00383111	$-0.08761790i$	$-1.297\times10^{20}i$	-0.0005366
		$N\tilde{A}_{H,2,0}^{-H,3,0}$	22.826344	0.99616889	0.08745023	1.297×10^{20}	0.0005366
L2 (1)	$N\tilde{a}_{1,1,0}^{1,2,0}$	$N\tilde{A}_{1,1,0}^{+1,2,0}$	$2.23629714i$	1.49987192	$-1.11786214i$	-1.693×10^{22}	-0.0700073
		$N\tilde{A}_{1,1,0}^{-1,2,0}$	1.73234665	0.50012808	0.88659144	1.693×10^{22}	0.0700073
	$N\tilde{a}_{1,2,0}^{1,3,0}$	$N\tilde{A}_{1,2,0}^{+1,3,0}$	$2.23630111i$	1.49986970	$-1.11785917i$	-1.693×10^{22}	-0.0700070
		$NA_{1,2,0}^{-1,3,0}$	1.73235177	0.50013030	0.86595016	1.693×10^{22}	0.0700070
	$N\tilde{a}_{1,3,0}^{1,4,0}$	$N\tilde{A}_{1,3,0}^{+1,4,0}$	$2.23630211i$	1.49986915	$-1.11785842i$	-1.693×10^{22}	-0.0700069
		$N\tilde{A}_{1,3,0}^{-1,4,0}$	1.73235306	0.50013085	0.86594984	1.693×10^{22}	0.0700069
L3 (2)	$N\tilde{a}_{2,1,0}^{2,2,0}$	$N\tilde{A}_{2,1,0}^{+2,2,0}$	$1.00484567i$	206.445991	$-206.443569i$	-6.957×10^{24}	-28.77280
		$N\tilde{A}_{2,1,0}^{-2,2,0}$	0.99512064	-204.44599	$204.443545i$	6.957×10^{24}	28.77280
	$N\tilde{a}_{2,2,0}^{2,3,0}$	$N\tilde{A}_{2,2,0}^{+2,3,0}$	1.00291622	343.40998	$-343.408528i$	-1.160×10^{25}	-47.95467
		$N\tilde{A}_{2,2,0}^{-2,3,0}$	0.99707525	-341.40998	$341.408520i$	1.160×10^{25}	47.95467
	$N\tilde{a}_{2,3,0}^{2,4,0}$	$N\tilde{A}_{2,3,0}^{+2,4,0}$	1.00208388	480.37398	$-480.372937i$	-1.160×10^{25}	-67.13654
		$N\tilde{A}_{2,3,0}^{-2,4,0}$	0.99791177	-478.37398	$478.372933i$	1.623×10^{25}	67.13654

Analysis of calculated data for excited lambda singulatons & ethertons - It follows from Tables 16.4.2 - 16.4.4 and Fig. 16.1.2:

- Negative and positive helyces making up excited singulatons and ethertons of the spacetime levels *L1* and *L2* are emitted in same directions, and they have opposite spins, similarly to the electrons and positrons. Their average translational velocities are equal to velocity of light and the frequencies are within the range of **gamma rays**.

- Negative and positive helyces making up excited singulatons and ethertons of the spacetime level *L3* are emitted in opposite directions, and they have opposite spins. Their translational velocities exceed significantly velocity of light and the frequencies exceed 10^{24} Hz. As the spacetime level becomes greater than *L3*, both speed and frequency increase.

- Combined ethertons and singulatons $c\widetilde{a}_{1,n,q}^{1,n,q}$ of the spacetime level *L2* ($m = 1$) decay according to the following sequence:

$$c\widetilde{a}_{1,3,0}^{1,4,0} \rightarrow c\widetilde{a}_{1,2,0}^{1,3,0} \rightarrow c\widetilde{a}_{1,1,0}^{1,2,0} \tag{16.4-1}$$

$$140.0146\ MeV \quad 140.0140\ MeV \quad 140.0138\ MeV$$

Notably, the energy of the combined ethertons and singulatons $c\widetilde{a}_{1,1,0}^{1,2,0}$ of 140.0138 *MeV* is about 3.7% greater than the measured energy 134.9766 *MeV* of the neutral pion π^0.

- Combined ethertons and singulatons $c\widetilde{a}_{m,n,q}^{m,n,q}$ of the spacetime level *L3* ($m = 2$) decay according to the following sequence:

$$c\widetilde{a}_{2,3,0}^{2,4,0} \rightarrow c\widetilde{a}_{2,2,0}^{2,3,0} \rightarrow c\widetilde{a}_{2,1,0}^{2,2,0} \tag{16.4-2}$$

$$134.273\ GeV \quad 95.909\ GeV \quad 57.546\ GeV$$

Notably, the energy of the combined ethertons and singulatons $c\widetilde{a}_{2,3,0}^{2,4,0}$ of 134.273 *GeV* is about 7.3% greater than the energy 125.09 *GeV* of the Higgs boson reported by CERN in 2017. The energy of the combined ethertons and singulatons $c\widetilde{a}_{2,2,0}^{2,3,0}$ of 95.909 *GeV* is about 5.1% greater than the energy 91.206 *GeV* of the Z boson reported by CERN in 2016.

16.5 Excited Harmonic Singulatons & Ethertons

Parental excited harmonic singulatrons - Table 16.5.1 shows compositions and properties of the parental harmonic singulatrons $\breve{a}_{H,n,q}^{0}$ of all spacetime levels *L*.

Table 16.5.1. Compositions and properties of parental excited harmonic singulatrons of all spacetime levels L with the quantum states $m = 0$.

Parental singula-trons	Toryces of parental excited harmonic singulatrons					
	Symbols	b_1	$\beta_1 = \beta_{2t}$	β_{2r}	δ_2	G
$\breve{a}^0_{H,0,0}$	$\breve{A}^{-2}_{H,0,0}$	-1.000000	-1.73205081i	2.000000	-1.000000	1.50000000
	$\breve{A}^{+2}_{H,0,0}$	0.333333	1.73205081i	-2.000000	3.000000	-1.50000000
$\breve{a}^0_{H,1,0}$	$\breve{A}^{-3}_{H,1,0}$	-0.500000	-2.82842712i	3.000000	-2.000000	2.66666667
	$\breve{A}^{+3}_{H,1,0}$	-0.250000	2.82842712i	-3.000000	4.000000	2.66666667
$\breve{a}^0_{H,2,0}$	$\breve{A}^{-3}_{H,2,0}$	-0.333333	-3.87298335i	4.000000	-3.000000	3.75000000
	$\breve{A}^{+3}_{H,2,0}$	0.200000	3.87298335i	-4.000000	5.000000	-3.75000000
$\breve{a}^0_{H,3,0}$	$\breve{A}^{-4}_{H,3,0}$	-0.250000	-4.89897949i	5.000000	-4.000000	4.80000000
	$\breve{A}^{+4}_{H,3,0}$	0.166667	4.89897949i	-5.000000	6.000000	-4.80000000

Emitted excited harmonic singulatons - Tables 16.5.2 shows compositions and properties of the emitted harmonic singulatons $\widetilde{\widetilde{a}}^{-H,n,q}_{H,n,q}$ of all spacetime levels L.

Table 16.5.2. Compositions and properties of emitted excited harmonic singulatons of all spacetime levels L with the quantum states $m = 0$.

Emitted singulatons	Helyces of emitted excited harmonic singulatons					
	Symbols	$\widetilde{b}_2$	$\widetilde{\beta}_{2t}$	$\widetilde{\beta}_{2r}$	$\widetilde{f}_2$, Hz	$\widetilde{E}_2$, MeV
$\widetilde{\widetilde{a}}^{H,1,0}_{H,0,0}$	$\widetilde{\widetilde{A}}^{-H,1,0}_{H,0,0}$	21.652662	0.99574321	0.09217080	1.441×10^{20}	0.596165
	$\widetilde{\widetilde{A}}^{+H,1,0}_{1,0,0}$	21.698796i	1.00425679	-0.09236719i	-1.441×10^{20}	-0.596165
$\widetilde{\widetilde{a}}^{H,2,0}_{H,1,0}$	$\widetilde{\widetilde{A}}^{-H,2,0}_{H,1,0}$	22.471741	0.99604727	0.08882477	1.339×10^{20}	0.553582
	$\widetilde{\widetilde{A}}^{+H,2,0}_{H,1,0}$	22.516197i	1.00395273	-0.08900050i	-1.339×10^{20}	-0.553582
$\widetilde{\widetilde{a}}^{H,3,0}_{H,2,0}$	$\widetilde{\widetilde{A}}^{-H,3,0}_{H,2,0}$	22.826344	0.99616999	0.08745023	1.297×10^{20}	0.536549
	$\widetilde{\widetilde{A}}^{+H,3,0}_{H,2,0}$	22.870111i	1.00383111	-0.08761790i	-1.297×10^{20}	-0.536549

Parental excited harmonic ethertrons - Table 16.5.3 shows compositions and properties of the parental excited harmonic ethertrons.

Table 16.5.3. Compositions and properties of parental excited harmonic ethertrons of all spacetime levels L with the quantum states $m = 0$.

Parental ethertrons	Symbols	Toryces of parental excited harmonic ethertrons				
		b_1	$\beta_1 = \beta_{2t}$	β_{2r}	δ_2	G
$a_{H,0,0}^0$	$A_{H,0,0}^{-1/2}$	21.652662	0.99574321	0.09217080	1.441×10^{20}	0.596165
	$A_{H,0,0}^{+1/2}$	0.66666667	0.86602540	-0.50000000	1.50000000	1.50000000
$a_{H,1,0}^0$	$A_{H,1,0}^{-1/3}$	1.50000000	0.94280904	0.33333333	0.66666667	-2.66666667
	$A_{H,1,0}^{+1/3}$	0.75000000	0.94280904	-0.33333333	1.33333333	2.66666667
$a_{H,2,0}^0$	$A_{H,2,0}^{-1/3}$	1.33333333	0.96824584	0.25000000	0.75000000	-3.75000000
	$A_{H,2,0}^{+1/3}$	0.80000000	0.96824584	-0.25000000	1.25000000	3.75000000
$a_{H,3,0}^0$	$A_{H,3,0}^{-1/4}$	1.25000000	0.97979690	0.20000000	0.80000000	-4.80000000
	$A_{H,3,0}^{+1/4}$	0.83333333	0.97979690	-0.20000000	1.20000000	4.80000000

Emitted excited harmonic ethertons - Table 16.5.4 shows compositions and properties of the excited harmonic ethertons $\widetilde{e}_{H,n,q}^{+H,n,q}$ of all spacetime levels L.

Table 16.5.4. Compositions and properties of emitted excited harmonic ethertons of all spacetime levels L with the quantum states $m = 0$.

Emitted ethertons	Symbols	Helyces of emitted excited harmonic ethertons				
		$\widetilde{b}_2$	$\widetilde{\beta}_{2t}$	$\widetilde{\beta}_{2r}$	$\widetilde{f}_2$, Hz	$\widetilde{E}_2$, MeV
$e_{H,0,0}^{+H,1,0}$	$\widetilde{E}_{H,0,0}^{+H,1,0}$	21.698796i	1.00425679	-0.09236719i	-1.441×10^{20}	-0.596165
	$\widetilde{E}_{1,0,0}^{+1,1,0}$	21.652662	0.99574321	0.09217080	1.441×10^{20}	0.596165
$e_{H,1,0}^{+H,2,0}$	$\widetilde{E}_{H,1,0}^{+H,2,0}$	22.516197i	1.00395273	-0.08900050i	-1.339×10^{20}	-0.553582
	$\widetilde{E}_{H,1,0}^{+H,2,0}$	22.471741	0.99604727	0.08882477	1.339×10^{20}	0.553582
$e_{H,2,0}^{+H,3,0}$	$\widetilde{E}_{H,2,0}^{+H,3,0}$	22.870111i	1.00383111	-0.08761790i	-1.297×10^{20}	-0.536549
	$\widetilde{E}_{H,2,0}^{+H,3,0}$	22.826344	0.99616999	0.08745023	1.297×10^{20}	0.536549

Analysis of calculated data for excited harmonic singulatons & ethertons - It follows from Tables 16.5.2 - 16.5.4 and Fig. 16.1.2:

- Real and imaginary helyces making up both excited singulatons and ethertons are emitted in the same direction, but they have opposite spins.
- The translational velocities of trailing strings $\widetilde{\beta}_{2t}$ of emitted real helyces making up excited singulatons and ethertons are slightly less than velocity of light, while these velocities of emitted imaginary helyces are slightly greater than velocity of light.

The calculated frequencies of electons and positons $\widetilde{f}_2$ are independent on the spacetime levels L and they are within the frequency range of the ***X-ray electromagnetic waves***.

16.6   Effect of Spacetime Levels of the Multiverse

Table 16.6 shows the effect of spacetime levels of the Multiverse on parameters of radiation particles emitted by their parental excited singulatron-ethertrons and electron-positrons making up a hydrogen atom. The data are applied to the case when the linear excitation quantum states n of these trons are reduced from $n = 2$ to $n = 1$.

Table 16.6. Effect of the spacetime levels L of the Multiverse on parameters of radiation particles emitted by constituent matter particles of a hydrogen atom when the linear excitation quantum states n of the trons are reduced from $n = 2$ to $n = 1$.

Space-time levels	Relative radii of emitted radiation particles $\widetilde{b}_2$		Relative velocities of emitted radiation particles $\widetilde{\beta}_{2t}$		Frequencies of emitted radiation particles $\widetilde{f}_2$, Hz		Energies of emitted radiation particles E_2, MeV	
	Singulaton-ethertons	Electon-positons	Singulaton-ethertons	Electon-positons	Singulaton-ethertons	Electon-positons	Singulaton-ethertons	Electon-positons
	ca-tons	**ce-tons**	**ca-tons**	**ce-tons**	**ca-tons**	**ce-tons**	**ca-tons**	**ce-tons**
L1	22.47174	5.47×10^2	1.00000	1.00000	1.339×10^{20}	2.261×10^{17}	0.0005536	1.87×10^{-3}
L2	1.984322	5.24×10^3	1.00000	1.00000	1.693×10^{22}	2.469×10^{15}	140.0146	2.04×10^{-5}
L3	0.999984	5.68×10^4	**205.446**	1.00000	6.957×10^{24}	2.102×10^{13}	5.754×10^4	1.74×10^{-7}
L4	≈ 1.0000	6.42×10^5	**65657.2**	1.00000	2.224×10^{27}	1.644×10^{11}	1.839×10^7	1.33×10^{-9}

As it follows from Table 16.6, as the spacetime levels of the Multiverse increases, the parameters of radiation particles change in the following manner:

- Radii of emitted singulaton-ethertons decrease
- Radii of emitted electon-positons increase

- Frequencies of emitted singulaton-ethertons increase
- Frequencies of emitted electon-positons decrease
- Energies of emitted singulaton-ethertons increase
- Energies of emitted electon-positons decrease
- Velocities of emitted singulaton-ethertons are equal to the velocity of light for the spacetime levels of the Multiverse *L1* and *L2*, while for higher spacetime levels their speeds exceed progressively the velocity of light as the spacetime level increases.
- Velocities of emitted electon-positons are equal to the velocity of light for all spacetime levels of the Multiverse.

<u>*Notes*</u>

17. Oscillated & Resonant Radiation Particles

In the proposed theory, the oscillated and resonant radiation particles are called ***trinos***. They are another form of radiation particles made up of polarized helyces. The trinos are emitted when either oscillation or resonant quantum states of their paternal electrons and positrons are reduced to the lower quantum states.

17.1 Oscillated Excited Negative Lambda Trinos

Negative and positive oscillated excited lambda trinos are emitted when the oscillation quantum states of their parental excited lambda electrons and positrons $e_{m,n,q}^{\pm 1}$ are reduced from higher to lower states as shown in Fig. 17.1. The emitted negative and positive trinos are unified and propagate alternatively along the same path forming neutral oscillated excited lambda trinos.

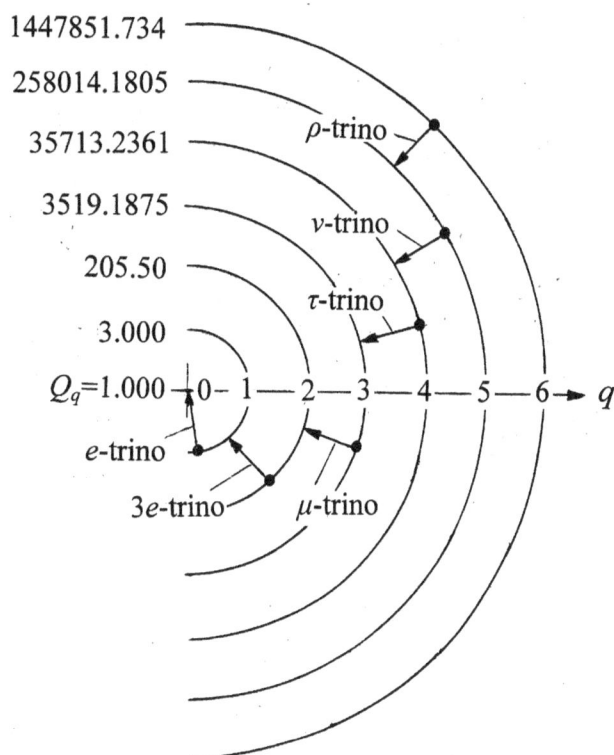

Fig. 17.1. Formation of oscillated excited lambda trinos.

Table 17.1.1 shows compositions and properties of the parental oscillated excited lambda electrons of the ordinary matter *L2* with $b_1 = 37538.0$ and $\breve{b}_1 = -37537.0$.

Table 17.1.1. Compositions and properties of parental oscillated excited lambda electrons of the ordinary matter *L2* with $b_1 = 37538.0$ and $\breve{b}_1 = -37537.0$.

q	Q_q	Parental trons	Toryces of parental trons				
			Symbols	$\beta_1 = \beta_{2t}$	β_{2r}	δ_2	G
0	1.00	$e_{2,1,0}^{-1}$	$E_{2,1,0}^{-}$	0.00729922	0.99997336	0.00002664	-5.328×10^{-5}
			$\breve{E}_{2,1,0}^{-}$	$-0.00729942i$	1.00002664	-0.00002664	5.328×10^{-5}
1	3.00	$e_{2,1,2}^{-1}$	$E_{2,1,1}^{-}$	0.00729922	0.99997336	0.00002664	-5.328×10^{-5}
			$\breve{E}_{2,1,1}^{-}$	$-0.00729942i$	1.00002664	-0.00002664	5.328×10^{-5}
2	205.5	$e_{2,1,2}^{-1}$	$Er_{2,1,2}^{-}$	0.00729922	0.99997336	0.00002664	-5.328×10^{-5}
			$\breve{E}r_{2,1,2}^{-}$	$-0.00729942i$	1.00002664	-0.00002664	5.328×10^{-5}
3	3519.1875	$e_{2,1,3}^{-1}$	$E_{2,1,3}^{-}$	0.00729922	0.99997336	0.00002664	-5.328×10^{-5}
			$\breve{E}_{2,1,3}^{-}$	$-0.00729942i$	1.00002664	-0.00002664	5.328×10^{-5}
4	35713.1236	$e_{2,1,4}^{-1}$	$E_{2,1,4}^{-}$	0.00729922	0.99997336	0.00002664	-5.328×10^{-5}
			$\breve{E}_{2,1,4}^{-}$	$-0.00729942i$	1.00002664	-0.00002664	5.328×10^{-5}
5	258014.181	$e_{2,1,5}^{-1}$	$E_{2,1,5}^{-}$	0.00729922	0.99997336	0.00002664	-5.328×10^{-5}
			$\breve{E}r_{2,1,5}^{-}$	$-0.00729942i$	1.00002664	-0.00002664	5.328×10^{-5}
6	1447851.73	$e_{2,1,6}^{-1}$	$E_{2,1,6}^{-}$	0.00729922	0.99997336	0.00002664	-5.328×10^{-5}
			$\breve{E}_{2,1,6}^{-}$	$-0.00729942i$	1.00002664	-0.00002664	5.328×10^{-5}
7	6642891.84	$e_{2,1,7}^{-1}$	$E_{2,1,7}^{-}$	0.00729922	0.99997336	0.00002664	-5.328×10^{-5}
			$\breve{E}_{2,1,7}^{-}$	$-0.00729942i$	1.00002664	-0.00002664	5.328×10^{-5}

Table 17.1.2 shows compositions and properties of oscillated excited negative trinos emitted by their parental oscillated excited lambda electrons of the ordinary matter *L2* with $b_1 = 37538.0$ and $\breve{b}_1 = -37537.0$.

Table 17.1.2. Compositions and properties of oscillated excited negative trinos emitted by their parental oscillated excited lambda electrons of the ordinary matter *L2* with $b_1 = 37538.0$ and $\breve{b}_1 = -37537.0$.

$q_k - q_j$	Emitted trino	Helyces of emitted trinos					
		Symbols	$\tilde{b}_2$	$\tilde{\beta}_{2t}$	$\tilde{\beta}_{2r}$	$\tilde{f}_2$, Hz	$\tilde{E}_2$, MeV
1 - 0	*e-trino* $\tilde{e}_{2,1,0}^{-2,1,1}$	$\tilde{E}_{2,1,0}^{-2,1,1}$	$2268.036693i$	1.00000039	$-0.00088182i$	-1.317×10^{16}	-0.0000545
		$\breve{\tilde{E}}_{2,1,0}^{-2,1,1}$	2268.036252	0.99999961	0.00088182	1.317×10^{16}	0.0000545
2 - 1	*3e-trino* $\tilde{e}_{2,1,1}^{-2,1,2}$	$\tilde{E}_{2,1,1}^{-2,1,2}$	$225.401495i$	1.00003937	$-0.00887323i$	-1.333×10^{18}	-0.0055133
		$\breve{\tilde{E}}_{2,1,1}^{-2,1,2}$	225.397058	0.99996063	0.00887306	1.333×10^{18}	0.0055133
3 - 2	*μ-trino* $\tilde{e}_{2,1,2}^{-2,1,3}$	$\tilde{E}_{2,1,2}^{-2,1,3}$	$55.728735i$	1.00064419	$-0.03589969i$	-2.181×10^{19}	-0.0902186
		$\breve{\tilde{E}}_{2,1,2}^{-2,1,3}$	55.710788	0.99935581	0.03588813	2.181×10^{19}	0.0902186
4 - 3	*τ-trino* $\tilde{e}_{2,1,3}^{-2,1,4}$	$\tilde{E}_{2,1,3}^{-2,1,4}$	$17.904232i$	1.00625858	$-0.11205499i$	-2.119×10^{20}	-0.8765516
		$\breve{\tilde{E}}_{2,1,3}^{-2,1,4}$	17.848292	-0.99374142	0.11170489	2.119×10^{20}	0.8765516
5 - 4	*v-trino* $\tilde{e}_{2,1,4}^{-2,1,5}$	$\tilde{E}_{2,1,4}^{-2,1,5}$	$6.876010i$	1.04321567	$-0.29715136i$	-1.463×10^{21}	-6.0523736
		$\breve{\tilde{E}}_{2,1,4}^{-2,1,5}$	6.729005	0.95678433	0.29079846	1.463×10^{21}	6.0523736
6 - 5	*ρ-trino* $\tilde{e}_{2,1,5}^{-2,1,6}$	$\tilde{E}_{2,1,5}^{-2,1,6}$	$3.105888i$	1.23130638	$-0.71841171i$	-7.833×10^{21}	-32.394560
		$\breve{\tilde{E}}_{2,1,5}^{-2,1,6}$	2.765238	0.76869362	0.63961716	7.833×10^{21}	32.394560
7 - 6	*χ-trino* $\tilde{e}_{2,1,6}^{-2,1,7}$	$\tilde{E}_{2,1,6}^{-2,1,7}$	$1.726368i$	2.00992433	$-1.74350103i$	-3.420×10^{22}	-141.44035
		$\breve{\tilde{E}}_{2,1,6}^{-2,1,7}$	0.990124	-0.00992433	0.99995075	-3.420×10^{22}	141.44035

17.2 Oscillated Excited Positive Lambda Trinos

Table 17.2.1 shows compositions and properties of the parental oscillated excited lambda positrons of the ordinary matter $L2$ with $b_1 = 0.50000666$ and $\breve{b}_1 = 0.49999334$.

Table 17.2.1. Compositions and properties of parental oscillated excited lambda positrons of the ordinary matter $L2$ with $b_1 = 0.50000666$ and $\breve{b}_1 = 0.49999334$.

q	Q_q	Parental trons	Toryces of parental positrons				
			Symbols	$\beta_1 = \beta_{2t}$	β_{2r}	δ_2	G
0	1.00	$e_{2,1,0}^{+1}$	$E_{2,1,0}^{+}$	0.00729922	-0.99997336	1.99997336	5.328×10^{-5}
			$\breve{E}_{2,1,0}^{+}$	$0.00729942i$	-1.00002664	2.00002664	-5.328×10^{-5}
1	3.00	$e_{2,1,2}^{+1}$	$E_{2,1,1}^{+}$	0.00729922	-0.99997336	1.99997336	5.328×10^{-5}
			$\breve{E}_{2,1,1}^{+}$	$0.00729942i$	-1.00002664	2.00002664	-5.328×10^{-5}
2	205.5	$e_{2,1,2}^{+1}$	$Er_{2,1,2}^{-}$	0.00729922	-0.99997336	1.99997336	5.328×10^{-5}
			$\breve{Er}_{2,1,2}^{-}$	$0.00729942i$	-1.00002664	2.00002664	-5.328×10^{-5}
3	3519.1875	$e_{2,1,3}^{+1}$	$E_{2,1,3}^{+}$	0.00729922	-0.99997336	1.99997336	5.328×10^{-5}
			$\breve{E}_{2,1,3}^{+}$	$0.00729942i$	-1.00002664	2.00002664	-5.328×10^{-5}
4	35713.1236	$e_{2,1,4}^{+1}$	$E_{2,1,4}^{+}$	0.00729922	-0.99997336	1.99997336	5.328×10^{-5}
			$\breve{E}_{2,1,4}^{+}$	$0.00729942i$	-1.00002664	2.00002664	-5.328×10^{-5}
5	258014.181	$e_{2,1,5}^{+1}$	$E_{2,1,5}^{+}$	0.00729922	-0.99997336	1.99997336	5.328×10^{-5}
			$\breve{Er}_{2,1,5}^{+}$	$0.00729942i$	-1.00002664	2.00002664	-5.328×10^{-5}
6	1447851.73	$e_{2,1,6}^{+1}$	$E_{2,1,6}^{+}$	0.00729922	-0.99997336	1.99997336	5.328×10^{-5}
			$\breve{E}_{2,1,6}^{+}$	$0.00729942i$	-1.00002664	2.00002664	-5.328×10^{-5}
7	6642891.84	$e_{2,1,7}^{+1}$	$E_{2,1,7}^{+}$	0.00729922	-0.99997336	1.99997336	5.328×10^{-5}
			$\breve{E}_{2,1,7}^{+}$	$0.00729942i$	-1.00002664	2.00002664	-5.328×10^{-5}

Table 17.2.2 shows compositions and properties of oscillated excited positive trinos emitted by their parental oscillated excited lambda positrons of the ordinary matter *L2* with $b_1 = 0.50000666$ and $\breve{b}_1 = 0.49999334$.

Table 17.2.2. Compositions and properties of oscillated excited positive trinos emitted by their parental oscillated excited lambda positrons of the ordinary matter *L2* with $b_1 = 0.50000666$ and $\breve{b}_1 = 0.49999334$.

$q_k - q_j$	Emitted trino	Helyces of emitted trinos					
		Symbols	$\tilde{b}_2$	$\tilde{\beta}_{2t}$	$\tilde{\beta}_{2r}$	$\tilde{f}_2$, Hz	$\tilde{E}_2$, MeV
1 - 0	$e+trino$ $\tilde{e}^{+2,1,1}_{2,1,0}$	$\tilde{E}^{+2,1,1}_{2,1,0}$	2268.036252	0.99999961	0.00088182	1.317×10^{16}	0.0000545
		$\breve{\tilde{E}}^{+2,1,1}_{2.1,0}$	2268.036693i	1.00000039	-0.00088182i	-1.317×10^{16}	-0.0000545
2 - 1	$3e+trino$ $\tilde{e}^{+2,1,2}_{2,1,1}$	$\tilde{E}^{+2,1,2}_{2,1,1}$	225.397058	0.99996063	0.00887306	1.333×10^{18}	0.0055133
		$\breve{\tilde{E}}^{+2,1,2}_{2.1,1}$	225.401495i	1.00003937	-0.00887323i	-1.333×10^{18}	-0.0055133
3 - 2	$\mu+trino$ $\tilde{e}^{+2,1,3}_{2,1,2}$	$\tilde{E}^{+2,1,3}_{2,1,2}$	55.710788	0.99935581	0.03588813	2.181×10^{19}	0.0902186
		$\breve{\tilde{E}}^{+2,1,3}_{2.1,2}$	55.728735i	1.00064419	-0.03589969i	-2.181×10^{19}	-0.0902186
4 - 3	$\tau+trino$ $\tilde{e}^{+2,1,4}_{2,1,3}$	$\tilde{E}^{+2,1,4}_{2,1,3}$	17.848292	-0.99374142	0.11170489	2.119×10^{20}	0.8765516
		$\breve{\tilde{E}}^{+2,1,4}_{2.1,3}$	17.904232i	1.00625858	-0.11205499i	-2.119×10^{20}	-0.8765516
5 - 4	$v+trino$ $\tilde{e}^{+2,1,5}_{2,1,4}$	$\tilde{E}^{+2,1,5}_{2,1,4}$	6.729005	0.95678433	0.29079846	1.463×10^{21}	6.0523736
		$\breve{\tilde{E}}^{+2,1,5}_{2.1,4}$	6.876010i	1.04321567	-0.29715136i	-1.463×10^{21}	-6.0523736
6 - 5	$\rho+trino$ $\tilde{e}^{+2,1,6}_{2,1,5}$	$\tilde{E}^{+2,1,6}_{2,1,5}$	2.765238	0.76869362	0.63961716	7.833×10^{21}	32.394560
		$\breve{\tilde{E}}^{+2,1,6}_{2.1,5}$	3.105888i	1.23130638	-0.71841171i	-7.833×10^{21}	-32.394560
7 - 6	$\chi+trino$ $\tilde{e}^{+2,1,7}_{2,1,6}$	$\tilde{E}^{+2,1,7}_{2,1,6}$	0.990124	-0.00992433	0.99995075	-3.420×10^{22}	141.44035
		$\breve{\tilde{E}}^{+2,1,7}_{2.1,6}$	1.726368i	2.00992433	-1.74350103i	-3.420×10^{22}	-141.44035

17.3 Analysis of Calculated Properties of Oscillated Excited Lambda Trinos

It follows from Tables 17.1.2 and 17.2.2:

- Real and imaginary helyces making up both negative and positive oscillated excited trinos are emitted in the same direction, but they have opposite spins.
- The translational velocities of trailing strings $\widetilde{\beta}_{2t}$ of emitted real helyces making up negative oscillated excited trinos are greater than velocity of light, while these velocities of emitted imaginary helyces are less than velocity of light.
- The translational velocities of trailing strings $\widetilde{\beta}_{2t}$ of emitted real helyces making up positive oscillated excited trinos are less than velocity of light, while these velocities of emitted imaginary helyces are greater than velocity of light.

The following changes of parameters of oscillated excited lambda trinos occur as the oscillation quantum states q increase:

- The relative radii $\widetilde{b}_2$ of their constituent helyces decrease
- The frequencies $\widetilde{f}_2$ of their constituent helyces increase
- The differences between the translational velocities of trailing strings $\widetilde{\beta}_{2t}$ of their real and imaginary helyces increase
- The energy of emitted trinos increase.

17.4 Resonant Harmonic Negative & Positive Trinos

Resonant harmonic trinos are emitted during decay of the unstable neutron $\downarrow nr_{L2}^{0}$. During that process, the resonant harmonic lambda electron $er_{0,0,0}^{-1}$ and positron $er_{0,0,0}^{+1}$ are respectively decayed into the harmonic lambda electron $e_{0,0,0}^{-1}$ and positron $e_{0,0,0}^{+1}$ (see Section 9.2).

Tables 17.4.1 and 17.4.2 show compositions and properties of parental harmonic electrons and positrons of resonant harmonic trinos.

Table 17.4.1. Compositions and properties of parental resonant harmonic electrons.

Parental trons	Toryces of parental electrons						
	Symbols	b_1	$\beta_1 = \beta_{2t}$	β_{2r}	δ_2	G
$e_{0,0,0}^{-1}$	$E_{0,0,0}^{-}$	2.00	0.86602540	0.50000000	0.50000000	-1.50000000
	$\breve{E}_{0,0,0}^{-}$	-1.00	-1.73205081i	2.00000000	-1.00000000	1.50000000
$er_{0,0,0}^{-1}$	$E_{0,0,0}^{-}$	2.00	0.86602540	0.50000000	0.50000000	-1.50000000
	$\breve{Er}_{0,0,0}^{-}$	-2.00	-1.11803399i	1.50000000	-0.50000000	0.83333333

Table 17.4.2. Compositions and properties of parental resonant harmonic positrons.

Parental trons	Toryces of parental positrons						
	Symbols	b_1	$\beta_1 = \beta_{2t}$	β_{2r}	δ_2	G
$e_{0,0,0}^{+1}$	$E_{0,0,0}^{+}$	0.66666667	0.86602540	0.50000000	0.50000000	-1.50000000
	$\breve{E}_{0,0,0}^{+}$	0.33333333	1.73205081i	-2.00000000	3.00000000	-1.50000000
$er_{0,0,0}^{+1}$	$E_{0,0,0}^{+}$	0.66666667	0.86602540	0.50000000	0.50000000	-1.50000000
	$\breve{Er}_{0,0,0}^{+}$	0.40000000	1.11803399i	-1.50000000	2.50000000	-0.83333333

Table 17.4.3 shows compositions and properties of the resonant harmonic trinos $\widetilde{er}_{0,0,0}^{0,0,0}$ emitted by the parental resonant harmonic electrons and positrons.

Table 17.4.3. Compositions and properties of resonant harmonic trinos emitted by the parental resonant harmonic electrons and positrons.

Emitted trino	Helyces of emitted trinos					
	Symbols	$\widetilde{b}_2$	$\widetilde{\beta}_{2t}$	$\widetilde{\beta}_{2r}$	$\widetilde{f}_2$, Hz	$\widetilde{E}_2$, MeV
$\widetilde{er}_{0,0,0}^{0,0,0}$	$\widetilde{\breve{Er}}_{0,0,0}^{-0,0,0}$	40.563927i	1.00121623	-0.04933487i	-4.12×10^{19}	-0.170333
	$\widetilde{Er}_{0,0,0}^{+0,0,0}$	40.539267	0.99878377	0.04930488	4.12×10^{19}	0.170333

Notes

SUMMARY OF THE BOOK

PART 1 - ABSTRACT MATHEMATICS OF A TORYX

1. TORYX BASIC STRUCTURE & PARAMETERS

Toryx is made up of a propagating double-circular leading string and a double-toroidal trailing string spiraling around and propagating synchronously with the leading string. Spacetime properties of a toryx are derived based on *three toryx spacetime postulates*. The first two postulates limit the degrees of freedom of its circular *leading string*, toroidal *trailing string* and its *eye radius*, while the third postulate requires the *spiral velocity of trailing string* to be constant and equal to the velocity of light.

The toryx spacetime postulates yield the *toryx law of planetary motion* for which the Kepler's third law of planetary motion is a particular case.

2. FEATURES OF ABSTRACT MATHEMATICS OF A TORYX

Based on the three toryx spacetime postulates, it is necessary to modify three commonly-accepted aspects of elementary mathematics and physics:

- Conventional zero (0) is replaced with *infinility* (± 0) that is an inverse of *infinity* ($\pm \infty$).
- Trigonometric cosine function between 180 and 360^0 is replaced with its inverse value.
- While spiral velocity of trailing string is constant and equal to the velocity of light, its either translational or rotational component *may exceed velocity of light*, causing the other component to be expressed with an imaginary number.

Toryces are topologically-polarized by their *vorticity* and *reality*:

- The *vorticity-polarized toryces* have opposite vorticities (negative and positive)
- The *reality-polarized toryces* have opposite realities (real and imaginary)
- Symmetrical relationships between topologically-polarized toryces are conveniently expressed by circular diagrams in which the unity (1) extends symmetrically towards both *infinity* ($\pm \infty$) and *infinility* (± 0).

3. CLASSIFICATION OF TORYCES

Based on their reality and vorticity, the toryces are divided into four main groups:

- Real negative toryces
- Real positive toryces
- Imaginary positive toryces
- Imaginary negative toryces.

Based on their ability to sustain their existence, the toryces are divided into two groups:

- *Mutually-sustained toryces* that sustain their existence only if these toryces are reality-polarized
- *Self-polarized toryces* that are capable to sustain their existence independently.

4. TRENDS OF TORYX PARAMETERS

Trends of spacetime properties of toryces are expressed as a function of steepness angle of toryx trailing string extending from 0^0 to 360^0 degrees, corresponding to the radius of toryx leading string extending from negative to positive infinity.

5. INVERSION OF TORYCES

As the radius of toryx leading string decreases from positive to negative infinity and the steepness angle of toryx trailing string φ_2 extending from 0^0 to 360^0 degrees, the toryx undergoes through four topological inversions:

- Toryx trailing string inverts at $\varphi_2 = 90^0$
- Wavelength of toryx trailing string and spherical boundary invert at $\varphi_2 = 180^0$
- Toryx leading string inverts at $\varphi_2 = 270^0$
- All toryx components invert at $\varphi_2 = 0^0 / 360^0$.

PART 2 - APPLIED MATHEMATICS OF A TORYX

6. QUANTUM STATES OF TORYCES

Toryces exist in two kinds of quantum spacetime states that further limit their degrees of freedom:

- *Excitation quantum states* at which only the radius of the toryx leading string increases, while its eye radius remains constant
- *Oscillation quantum states* at which both the radius of the toryx leading string and its eye radius change in the same proportion.

Toryces in which their frequencies of trailing strings relate to one another by simple harmonic ratios are called *harmonic toryces*. The harmonic toryces in which the frequencies of toryx trailing strings relate to one another by the harmonic ratios 1:3:5 are called *resonant harmonic toryces*. Toryces which quantum states are a function of the golden ratio ϕ are called *golden toryces*.

The derived quantization equations for excited toryces are based on three proposed limitations of degrees of freedom of real negative *lambda, harmonic* and *golden toryces* shown in Exhibit 6.1.

Exhibit 6.1. Limitations of degrees of freedom of excited toryces.

The relative radius of leading string of real negative toryx b_1 is equal to:		
Lambda toryx	**Harmonic toryx**	**Golden toryx**
$b_1 = z = 2(n\Lambda)^m$ (6.1-1)	$b_1 = z = 2 + n$ (6.1-2)	$b_1 = z = 2 + n/\phi$ (6.1-3)

In Exhibit 6.1:

z = toryx quantization parameter
$m \rightarrow 0, 1, 2, . . $, toryx quantum states dependent on the spacetime levels of the Multiverse
$n \rightarrow 0, 1, 2, . . $, toryx linear excitation quantum states
$\Lambda = 137, ...$, *toryx quantization constant*.

The proposed quantization equation for the relative radii b_1 of toryces making up excited electrons of atomic electrons in a hydrogen atom was verified based on the experimental data for the orbital radii of atomic electrons in a hydrogen atom. It was then used for derivation of quantization equations for other excited toryces making up positrons, ethertons and singulatons. Another equation was proposed for calculation of quantum states of oscillated toryces. This equation was calibrated based on the experimental data for leptons.

7. FORMATION OF ELEMENTARY MATTER PARTICLES

Toryces are created spontaneously from *quantum vacuum* in a form of harmonic toryces. The process is governed by the *toryx uncertainty principle*.

Four kinds of basic elementary matter particles, called *trons*, are formed by the unification of polarized toryces:

- *Electrons* that model known electrons
- *Positrons* that model known positrons
- *Ethertrons* that model still unknown light elementary particles
- *Singulatrons* that model known neutral pions.

Spacetime properties of toryces and elementary matter particles are expressed by using the *objective spacetime units* comprehendible by all intelligent inhabitants in the Universe. Subjective *physical properties of toryces* invented by human beings are directly related to their respective *spacetime properties*. Stable trons follow the *tron polarization conservation law* assuring a balanced absorption and release of spacetime by their constituent toryces.

All four basic elementary matter particles may exist at various spacetime levels L, besides the spacetime level $L2$ corresponding to the *ordinary matter* as shown in Table 7.4. The exponential excitation quantum states m of toryces forming trons depend on the spacetime levels L of the Multiverse to which they belong.

Table 7.4. Spacetime levels of the Multiverse.

Spacetime levels	Excitation quantum states m			
	Electron	**Positron**	**Singulatron**	**Ethertron**
L0	$m = 0$	$m = 0$	$m = 0$	$m = 0$
L1	$m = 1$	$m = 1$	$m = 0$	$m = 0$
L2	**$m = 2$**	**$m = 2$**	**$m = 1$**	**$m = 1$**
L3	$m = 3$	$m = 3$	$m = 2$	$m = 2$
L4	$m = 4$	$m = 4$	$m = 3$	$m = 3$
Lx	$m = x$	$m = x$	$m = x - 1$	$m = x - 1$
....	....	....	....	....

L2 = spacetime level of the ordinary matter.

8. EXAMPLES OF ELEMENTARY MATTER PARTICLES

This chapter provides examples of structures and properties of various elementary matter particles and their constituent toryces.

9. NUCLEONS, LIGHT ATOMS & ISOTOPES

Nucleons, atom and isotopes are the assemblies of elementary matter particles:

Nucleon crystal – It is made up of a bi-pyramid hexagonal crystal structure formed at the intersections of harmonic toryces. It could be either unexcited or excited.

Nucleon core - It contains exciting ethertrons and singulatrons residing at the vertices and a center of the nucleon crystal.

Unstable neutron – It contains an excited nucleon crystal, nucleon core and resonant oscillated harmonic electron and positron located at the center of the nucleon crystal.

Proton - It is made up of an unexcited nucleon crystal, a nucleon core, a harmonic positron and a harmonic oscillated electron located at the center of the nucleon crystal.

Stable neutron – It is composed of a proton and a harmonic electron located at the center of the proton.

Hydrogen atom is made up of an unexcited nucleon crystal, a nucleon core, an excited positron and an excited electron.

Nucleons of stable atoms and *isotopes* - They are made up of protons and stable neutrons.

Nucleons of unstable atoms and *isotopes* – They contain protons and neutrons with at least one of neutrons being unstable.

10. OSCILLATED ELEMENTARY PARTICLES

This chapter shows calculated properties of oscillated elementary particles, including:

7 leptons (see Table 10.1), 12 quarks (see Table 10.2), and 24 mesons (see Tables 10.3.1 – 10.3.8).

It provides a comparison between calculated and measured values of their relative magnetic moments and masses.

11. TORYCES OF THE MACRO-WORLD

Previous chapters described a role of the *micro-toryces* in the formation of matter particles and atoms of the micro-world. In the macro-world, the assemblies of atoms contained in each body form the macro-spacetimes called the *macro-toryces* that become integral parts of each body.

Macro-toryces are the spacetime fields associated with celestial bodies and responsible for the interactions between them. Equations describing spacetime properties of micro- and macro-toryces are the same, except for the equations describing their eye radii.

The macro-toryces obey two laws:

- The *macro-toryx law of planetary motion* that for large distances between celestial bodies reduces to the Kepler's third law of planetary motion
- The *spacetime law of gravitation* that for large distances between celestial bodies reduces to the Newton's universal law of gravitation.

PART 3 - ABSTRACT & APPLIED MATHEMATICS OF A HELYX

12. HELYX BASIC STRUCTURE & PARAMETERS

Helyx is single-level *helicola* containing a straight-line leading string and a helical trailing string propagating synchronously with the leading string. Spacetime properties of a helyx are derived based on *three helyx spacetime postulates*. The first two postulates limit the degrees of freedom of its straight-line *leading string*, helical *trailing string* and its *eye radius*, while the third postulate requires the *spiral velocity of trailing string* to be constant and equal to the velocity of light.

13. FEATURES OF ABSTRACT MATHEMATICS OF A HELYX

Based on the three helyx spacetime postulates, it is necessary to modify three commonly-accepted aspects of elementary mathematics and physics:

- Conventional zero (0) is replaced with *infinility* (± 0) that is an inverse of *infinity* ($\pm\infty$).
- Trigonometric cosine function between 180 and 360^0 is replaced with its inverse value.
- While spiral velocity of trailing string is constant and equal to the velocity of light, its either translational or rotational component *may exceed velocity of light*, causing the other component to be expressed with an imaginary number.

Helyces are topologically-polarized by their *vorticity* and *reality*:

- The *vorticity-polarized helyces* have opposite vorticities (negative and positive)
- The *reality-polarized helyces* have opposite realities (real and imaginary)
- Symmetrical relationships between topologically-polarized helyces are conveniently expressed by circular diagrams in which the unity (1) extends symmetrically towards both *infinity* ($\pm\infty$) and *infinility* (± 0).

14. CLASSIFICATION OF HELYCES

Based on their reality and vorticity, the helyces are divided into four main groups:

- Real negative helyces
- Real positive helyces
- Imaginary positive helyces
- Imaginary negative helyces.

Based on their ability to sustain their existence, the helyces are divided into two groups:

- *Mutually-sustained helyces* that sustain their existence only if these toryces are reality-polarized
- *Self-polarized helyces* that are capable to sustain their existence independently.

15. TRENDS & INVERSIONS OF HELYX PARAMETERS

Trends of spacetime properties of toryces are expressed as a function of apex angle of helyx trailing string extending from 0^0 to 360^0 degrees, corresponding to the radius of helyx leading string extending from negative to positive infinity.

As the radius of helyx leading string decreases from positive to negative infinity and the apex angle of helyx trailing string $\tilde{\varphi}_2$ extending from 0^0 to 360^0 degrees, the helyx undergoes through four topological inversions:

- Real negative helyces - ***Wavelength of trailing string*** becomes inverted at $\widetilde{\varphi}_2 = 90^0$
- Real positive helyces - ***Trailing string*** becomes inverted at $\widetilde{\varphi}_2 = 180^0$
- Imaginary positive - ***Spiral length of trailing string*** becomes inverted at $\widetilde{\varphi}_2 = 270^0$
- Imaginary negative - ***Entire helyx becomes inverted*** when $\widetilde{\varphi}_2 = 0^0 / 360^0$.

16. BASIC ELEMENTARY RADIATION PARTICLES

Elementary radiation particles are formed when the quantum states m, n and q of their paternal trons and toryces are reduced to the lower levels. The frequencies of emitted helyces are defined based on the ***spacetime intensity conservation law***.

The elementary radiation particles emitted by the excited trons are called ***tons*** as shown in Table 16.1.

Table 16.1. Types of elementary radiation particles.

Parental matter particles		Elementary radiation particles	
Combined	Constituent	Constituent	Combined
ce-tron	Electron	Electon	ce-ton
	Positron	Positon	
ca-tron	Ethertron	Etherton	ca-ton
	Singulatron	Singulaton	

Two basic radiation particles are emitted jointly by two combined parental trons: the ce-tron $ce_{m,n,q}^{0}$ and the ca-tron $ca_{m,n,q}^{0}$. They emit respectively the combined electron-positron ce-tons $c\widetilde{e}_{m,n,q}^{m,n,q}$ (Fig. 16.1.1) and the etherton-singulatons ca-trons $c\widetilde{a}_{m,n,q}^{m,n,q}$ (Fig. 16.1.2).

Similarly to the trons, there are two kinds of tons, the ***charge-polarized tons*** and the ***reality-polarized tons***. Both of them are formed when their respective polarized parental trons are transferred from higher to lower quantum states.

The radiation particles emitted by the ce-tons of the spacetime levels *L1* and *L2* (ordinary matter) propagate with velocity of light, while the radiation particles emitted by the ca-tons of the spacetime levels *L3* and higher propagate with superluminal velocities.

The combined ca-ton $c\widetilde{a}_{1,n,q}^{1,n,q}$ of the spacetime level *L2* (m = 1) decays according to the following sequence:

$$c\tilde{a}_{1,3,0}^{1,4,0} \rightarrow c\tilde{a}_{1,2,0}^{1,3,0} \rightarrow c\tilde{a}_{1,1,0}^{1,2,0}$$

$$140.0146 \; MeV \quad 140.0140 \; MeV \quad 140.0138 \; MeV$$

The energy of a combined ce-ton $c\tilde{a}_{1,1,0}^{1,2,0}$ of $140.0138 \; MeV$ is about 3.7% greater than the measured energy $134.9766 \; MeV$ of the neutral pion π^0.

The combined ca-ton $c\tilde{a}_{m,n,q}^{m,n,q}$ of the spacetime level $L3$ ($m = 2$) decays according to the following sequence:

$$c\tilde{a}_{2,3,0}^{2,4,0} \rightarrow c\tilde{a}_{2,2,0}^{2,3,0} \rightarrow c\tilde{a}_{2,1,0}^{2,2,0}$$

$$134.273 \; GeV \quad 95.909 \; GeV \quad 57.546 \; GeV$$

Notably, the energy of the combined ca-ton $c\tilde{a}_{2,3,0}^{2,4,0}$ of $134.273 \; GeV$ is about 7.3% greater than the energy $125.09 \; GeV$ of the Higgs boson reported by CERN in 2017. The energy of the combined ethertons and singulatons $c\tilde{a}_{2,2,0}^{2,3,0}$ of $95.909 \; GeV$ is about 5.1% greater than the energy $91.206 \; GeV$ of the Z boson reported by CERN in 2016.

17. OSCILLATED & RESONANT RADIATION PARTICLES

The oscillated and resonant radiation particles are called *trinos*. They are used to model neutrinos that are currently thought to be the matter particles. Negative and positive oscillated excited trinos are emitted when the oscillation quantum states of their parental oscillated excited electrons and positrons are reduced from higher to lower states. The negative and positive helyces of emitted oscillated trinos are unified and propagate alternatively along the same path, forming neutral oscillated trinos. Negative and positive resonant harmonic trinos are emitted during decay of unstable neutrons when the resonant quantum states of their parental electrons and positrons are reduced from higher to lower states. The emitted negative and positive helyces of emitted resonant trinos are unified and propagate alternatively along the same path, forming neutral resonant trinos.

Comparisons with Measured Data

The values of calculated gravitational masses, charges and magnetic moments of several known particles, atoms and isotopes that are close to their measured values:

- Hydrogen atom (see Table 9.4.1)
- Proton (see Table 9.3.2)
- Unstable neutron (see Table 9.3.1)

- Deuterium (see Table 9.4.2)
- Tritium (see Table 9.4.3)
- Helium-3 (see Table 9.4.4)
- Helium-4 (see Table 9.4.5)
- Electron, muon and tau (see Table 10.1 and Fig. 10.1)
- Basic quarks (see Table 10.2)
- Mesons (see Tables 10.3.1 – 10.3.8).

Table 16.2.5 shows a comparison of the calculated frequencies $\tilde{f}_2$ of emitted ce-tons with frequencies of spectra lines for hydrogen atom calculated from the Rydberg's equation.

Table 16.2.5. Comparison of the calculated frequencies $\tilde{f}$ of emitted ce-tons with frequencies of spectra lines for hydrogen atom calculated from the Rydberg's Eq. (16.2-1).

Spectra lines Of hydrogen	Quantum states $(k-j)$	Excited ce-tons	Frequencies $\tilde{f}_2$, Hz		Calculated/ Rydbers's
			Calculated	Rydberg's	
H_α	$(3-2)$	$c\tilde{e}_{2,2,0}^{2,3,0}$	4.571648×10^{14}	4.569225×10^{14}	1.000530
H_β	$(4-2)$	$c\tilde{e}_{2,2,0}^{2,4,0}$	6.171721×10^{14}	6.168454×10^{14}	1.000530
H_γ	$(5-2)$	$c\tilde{e}_{2,2,0}^{2,5,0}$	6.912325×10^{14}	6.908668×10^{14}	1.000529
H_δ	$(6-2)$	$c\tilde{e}_{2,2,0}^{2,6,0}$	7.314629×10^{14}	7.310760×10^{14}	1.000529

Predictions

a) Stable neutron & leptons with predicted structures and properties:

- Stable neutron (see Table 9.3.3)
- Leptons (see Table 10.1 and Fig. 10.1):

> *3e* lepton............1.5199893 MeV
> *Nu* (v) lepton.........18.2494 GeV
> *Rho* (ρ) ....131.8450 GeV
> *Chi* (χ) lepton......739.8507 GeV.

Notably, the calculated mass of **rho-tron** is about 5.4% greater than the mass of the Higgs boson of 125.09 GeV measured at CERN in 2017.

b) Stable elementary matter particles with predicted structures and properties:

- Reality-polarized harmonic lambda electron (see Table 8.1)
- Reality-polarized harmonic lambda positron (see Table 8.2)
- Vorticity-polarized resonant harmonic lambda ethertron (see Table 8.3)
- Vorticity-polarized resonant harmonic lambda singulatron (see Table 8.4)
- Self-polarized unexcited harmonic electron (see Table 8.5)
- Self-polarized unexcited harmonic positron (see Table 8.6)
- Self-polarized excited harmonic electron (see Table 8.7)
- Self-polarized excited harmonic positron (see Table 8.8)
- Excited lambda electron (See Table 8.9)
- Excited lambda positron (See Table 8.10)
- Excited lambda ethertron (See Table 8.11)
- Excited lambda singulatron (See Table 8.12)
- Combined harmonic lambda ce-tron (See Table 8.13)
- Combined harmonic lambda ca-tron (See Table 8.14)
- Combined excited lambda ce-tron (See Table 8.15)
- Combined excited lambda ca-tron (See Table 8.16)
- Reality-polarized oscillated resonant harmonic lambda electron (see Table 8.17).

c) Unstable matter particles with predicted structures and properties:

- Quarks (see Table 10.2)
- Mesons (see Tables 10.3.1 – 10.3.8)

d) Stable elementary radiation particles with predicted structures and properties for different spacetime levels:

- Excited lambda electons (see Table 16.2.2)
- Excited lambda positons (see Table 16.2.4)
- Excited harmonic electons (see Table 16.3.2)
- Excited harmonic positons (see Table 16.3.4)
- Excited lambda singulatons (see Table 16.4.2)
- Excited lambda ethertons (see Table 16.4.4)
- Excited harmonic singulatons (see Table 16.5.2)
- Excited harmonic ethertons (see Table 16.5.4).

e) Oscillated and resonant radiation particles with predicted structures and properties:

- Oscillated excited negative trinos (see Tables 17.1.2)
- Oscillated excited positive trinos (see Tables 17.2.2)
- Resonant harmonic trinos (see Table 17.4.3).

f) Predicted effect of spacetime levels on properties of hydrogen atoms & protons:

Both the compositions and properties of hydrogen atoms and protons are greatly dependent on their spacetime levels L of the Multiverse.

Table 9.5 compares several physical properties of hydrogen atoms of the spacetime levels $L1$, $L2$ and $L3$ of the Multiverse.

Table 9.5. Relative parameters of the hydrogen atoms $\downarrow H_L^0$
in the ground state $n = 1$ of three spacetime levels L of the Multiverse.

Spacetime levels L of the Multiverse	Hydrogen atoms	Orbital electron radius ratio	Orbital electron magnetic moment ratio	Hydrogen atom mass ratio	Modulus of elasticity ratio
$L1$	$\downarrow H_{L1}^0$	1/137.25	1/11.37	1/37.08	3.53×10^8
L2 (ordinary matter)	$\downarrow H_{L2}^0$	**1.0**	**1.0**	**1.0**	**1.0**
$L3$	$\downarrow H_{L3}^0$	137.00	11.70	136.30	3.52×10^{-8}

As it follows from Table 9.5, as the spacetime level L of the Multiverse increases the orbital electron radius, the orbital electron magnetic moment and the mass of the hydrogen atom increase, while its modulus of elasticity decreases.

Table 9.6 compares several physical properties of protons of the spacetime levels $L1$, $L2$ and $L3$ of the Multiverse.

Table 9.6. Physical properties of the protons p_L^{+1}
in the ground state $n = 1$ of three spacetime levels L of the Multiverse.

Spacetime levels L	Proton	Proton magnetic moment ratio	Proton mass ratio	Calculated mass in respect to measured mass* GeV/c^2
$L1$	$\downarrow p_{L1}^{+1}$	- 0.45533	1/37.78	0.024835152
L2 (ordinary matter)	$\uparrow p_{L2}^{+1}$	**1.0**	**1.0**	**0.938272046***
$L3$	$\uparrow p_{L3}^{+1}$	0.99986	136.365	127.9474747

As it follows from Table 9.6, as the spacetime level L of the Multiverse increases the proton magnetic moment increases and reaches its maximum value for the spacetime level $L2$. It is then decreases slightly as the spacetime level L increases. The proton mass increases with the increase of the spacetime level L.

The right column of Table 9.6 shows calculated proton masses of the spacetime levels *L1* and *L3* in respect to the measured proton mass* of the spacetime levels *L2* based on the proton mass ratios shown in the previous column. Notably, for the spacetime level *L3*, the calculated proton mass is equal to 127.9474747 GeV/c^2 that is 2.28% greater than the mass 125.09 ± 0.24 GeV/c^2 of the Higgs boson reported by CERN in 2017.

Table 16.6. Effect of the spacetime levels L of the Multiverse on parameters of radiation particles emitted by constituent matter particles of hydrogen atom when the linear excitation quantum states n of the trons are reduced from $n = 2$ to $n = 1$.

Space-time Levels	Relative radii of emitted radiation particles $\tilde{b}_2$		Relative velocities of emitted radiation particles $\tilde{\beta}_{2t}$		Frequencies of emitted radiation particles $\tilde{f}_2$, Hz		Energies of emitted radiation particles $\tilde{E}_2$, MeV	
	Singula-ton-ethertons	Electon-positons	Singula-ton-ethertons	Electon-positons	Singula-ton-ethertons	Electon-positons	Singula-ton-ethertons	Electon-positons
	ca-tons	*ce-tons*	*ca-tons*	*ce-tons*	*ca-tons*	*ce-tons*	*ca-tons*	*ce-tons*
L1	22.47174	5.47×10^2	1.00000	1.00000	1.339×10^{20}	2.261×10^{17}	0.0005536	1.87×10^{-3}
L2	1.984322	5.24×10^3	1.00000	1.00000	1.693×10^{22}	2.469×10^{15}	140.0146	2.04×10^{-5}
L3	0.999984	5.68×10^4	**205.446**	1.00000	6.957×10^{24}	2.102×10^{13}	5.754×10^4	1.74×10^{-7}
L4	≈ 1.0000	6.42×10^5	**65657.2**	1.00000	2.224×10^{27}	1.644×10^{11}	1.839×10^7	1.33×10^{-9}

As it follows from Table 16.6, as the spacetime levels of the Multiverse increases, the parameters of radiation particles change in the following manner:

- Radii of emitted singulaton-ethertons decrease
- Radii of emitted electon-positons increase
- Frequencies of emitted singulaton-ethertons increase
- Frequencies of emitted electon-positons decrease
- Energies of emitted singulaton-ethertons increase
- Energies of emitted electon-positons decrease
- Velocities of emitted singulaton-ethertons are equal to the velocity of light for spacetime levels of the Multiverse *L1* and *L2*, while for higher spacetime levels their speeds exceed progressively the velocity of light as the spacetime level increases.
- Velocities of emitted electon-positons are equal to the velocity of light for all spacetime levels of the Multiverse.

<u>*Notes*</u>

Notes

Appendix A

DERIVATION OF TORYX EQUATIONS

CONTENTS

A1 Toryx Parameters as Functions of b_1

A1.1 Toryx Radii

Eq. (1.6-1a)
Relative radius of toryx leading string – From Eq. (1.4-2):

$$b_1 = \frac{r_1}{\tilde{r_0}}$$

(A1.1-1)

Eq. (1.6-1b)
Relative radius of toryx trailing string – From Eqs. (1.4-3) and (1.5-2):

$$b_2 = \frac{r_2}{r_0} = b_1 - 1$$

(A1.1-2)

Eq. (1.6-13)
Relative radius of toryx spherical boundary – From Eqs. (A1.1-1) and Fig. 1.1.2:

$$b = \frac{r}{r_0} = \frac{2r_1 - r_0}{r_0} = 2b_1 - 1$$

$$b = 2b_1 - 1$$

(A1.1-3)

Eq. (6.5-2)
Relative radius of toryx fine-structure spherical boundary – From Eqs. (6.5-1), (A1.1-1) and Fig. (6.5.1):

$$b_3 = \frac{r_3}{r_0} = \frac{\sqrt{r_1^2 - r_2^2}}{r_0} = \frac{\sqrt{r_1^2 - (r_1 - r_0)^2}}{r_0} = \frac{\sqrt{2r_1 - r_0}}{r_0} = \sqrt{2b_1 - 1}$$

$$b_3 = \sqrt{2b_1 - 1} \tag{A1.1-4}$$

A1.2 Toryx Lengths, Wavelengths & Number of Windings

Eq. (1.6-3a)
Relative length of one winding of toryx leading string – From Eqs. (1.4-4) and (A1.1-1):

$$l_1 = \frac{L_1}{2\pi r_0} = \frac{2\pi r_1}{2\pi r_0} = b_1$$

$$l_1 = \frac{L_1}{2\pi r_0} = b_1 \tag{A1.2-1}$$

Eq. (1.6-3b)
Relative length of one winding of toryx trailing string – From Eqs. (1.3-1) and (1.4-5):

$$l_2 = \frac{L_2}{2\pi r_0} = \frac{2\pi r_1}{2\pi r_0} = b_1$$

$$l_2 = \frac{L_2}{2\pi r_0} = b_1 \tag{A1.2-2}$$

Eq. (1.6-2b)
Relative wavelength of toryx trailing string – From Eqs. (1.4-17), (A1.1-1), (A1.1-2) and Fig. 1.3:

$$\eta_2 = \frac{\lambda_2}{2\pi r_0} = \frac{\sqrt{(2\pi r_1)^2 - (2\pi r_2)^2}}{2\pi r_0} = \sqrt{b_1^2 - b_2^2} = \sqrt{b_1^2 - (b_1-1)^2} = \sqrt{2b_1 - 1}$$

$$\eta_2 = \frac{\lambda_2}{2\pi r_0} = \sqrt{2b_1 - 1} \tag{A1.2-3}$$

Eq. (1.6-5b)
The number of widings of toryx trailing string – From Eqs. (A1.2-1) and (A1.2-3):

$$w_2 = \frac{L_1}{\lambda_2} = \frac{2\pi b_1 r_0}{2\pi r_0 \sqrt{2b_1 - 1}} = \frac{b_1}{\sqrt{2b_1 - 1}}$$

$$w_2 = \frac{b_1}{\sqrt{2b_1 - 1}} \tag{A1.2-4}$$

A1.3 Steepness Angle of Trailing String

Eq. (1.6-4b)
- **Cosine of steepness angle of toryx trailing string** – From Eqs. (1.6-11), (1.6-12) and Fig. 1.5:

$$\cos s\varphi_2 = \frac{b_1-1}{b_1} \tag{A1.3-1}$$

- **Sine of steepness angle of toryx trailing string** – From Eq. (A1.3-1):

$$\sin s\varphi_2 = \sqrt{1-\cos s^2\varphi_2} = \sqrt{1-\left(\frac{b_1-1}{b_1}\right)^2} = \frac{\sqrt{2b_1-1}}{b_1}$$

$$\sin s\varphi_2 = \frac{\sqrt{2b_1-1}}{b_1} \tag{A1.3-2}$$

- **Tangent of steepness angle of toryx trailing string** – From Eqs. (A1.3-1) and (A1.3-2):

$$\tan s\varphi_2 = \frac{\sin s\varphi_2}{\cos s\varphi_2} = \frac{\sqrt{2b_1-1}}{b_1}\frac{b_1}{b_1-1} = \frac{\sqrt{2b_1-1}}{b_1-1}$$

$$\tan s\varphi_2 = \frac{\sqrt{2b_1-1}}{b_1-1} \tag{A1.3-3}$$

A1.4 Middle Velocities of Toryx Strings (Fig. 1.1.3, point *a*)

Eq. (1.6-7b)
Relative rotational velocity of toryx trailing string – From Eqs. (1.4-13), (1.5-3), (A1.3-1) and Fig. 1.5:

$$\beta_{2r} = \frac{V_{2r}}{c} = \beta_2 \cos s\varphi_2 = \frac{b_1-1}{b_1}$$

$$\beta_{2r} = \frac{V_{2r}}{c} = \frac{b_1-1}{b_1} \tag{A1.4-1}$$

Eq. (1.6-6b)
Relative translational velocity of toryx trailing string – From Eqs. (1.4-12), (1.5-3), (A1.4-1), and Fig. 1.5:

$$\beta_{2t} = \frac{V_{2t}}{c} = \sqrt{1 - \beta_{2r}^2} = \sqrt{1 - \left(\frac{b_1 - 1}{b_1}\right)^2} = \frac{\sqrt{2b_1 - 1}}{b_1}$$

$$\beta_{2t} = \frac{V_{2t}}{c} = \frac{\sqrt{2b_1 - 1}}{b_1} \tag{A1.4-2}$$

Eqs. (1.6-7a) and (1.6-8a)
Relative spiral and rotational velocities of toryx leading string – From Eqs. (1.1-4), (1.4-8), (1.4-10) and (A1.4-2):

$$\beta_1 = \beta_{1r} = \beta_{2t} = \frac{\sqrt{2b_1 - 1}}{b_1}$$

$$\beta_1 = \frac{V_1}{c} = \beta_{1r} = \frac{V_{1r}}{c} = \frac{\sqrt{2b_1 - 1}}{b_1} \tag{A1.4-3}$$

A1.5 Frequencies of Toryx Strings

Eq. (1.6-9a)
Relative frequency of toryx leading string – From Eqs. (1.4-14), (1.4-18), (A1.2-1) and (A1.4-3):

$$\delta_1 = \frac{f_1}{f_0} = \frac{V_1}{L_1 f_0} = \frac{V_1}{c} \frac{c}{2\pi r_0} \frac{1}{b_1 f_0} = \frac{\beta_1}{b_1} = \frac{\sqrt{2b_1 - 1}}{b_1^2}$$

$$\delta_1 = \frac{f_1}{f_0} = \frac{\sqrt{2b_1 - 1}}{b_1^2} \tag{A1.5-1}$$

Eq. (1.6-9b)
Relative frequency of toryx trailing string – From Eqs. (1.3-1), (1.5-3), (1.4-11), (1.4-18) and (A1.2-2):

$$\delta_2 = \frac{f_2}{f_0} = \frac{V_2}{L_2 f_0} = \frac{c}{2\pi r_0} \frac{\beta_2}{b_1 f_0} = \frac{1}{b_1}$$

$$\delta_2 = \frac{f_2}{f_0} = \frac{1}{b_1} \tag{A1.5-2}$$

A1.6 Peripheral Velocities of Trailing String (Fig. 3.3.1)

Eq. (3.3-1)
Relative inner translational velocity of trailing string - From Eq. (A1.4-2):

$$\beta_{2t}^{in} = \beta_{2t} \frac{1}{b_1} = \frac{\sqrt{2b_1 - 1}}{b_1^2}$$

$$\beta_{2t}^{in} = \frac{\sqrt{2b_1 - 1}}{b_1^2} \tag{A1.6-1}$$

Eq. (3.3-2)
Relative outer translational velocity of trailing string - From Eq. (A1.4-2):

$$\beta_{2t}^{out} = \beta_{2t} \frac{2b_1 - 1}{b_1} = \frac{\sqrt{2b_1 - 1}(2b_1 - 1)}{b_1^2} = \frac{(2b_1 - 1)^{1.5}}{b_1^2}$$

$$\beta_{2t}^{out} = \frac{(2b_1 - 1)^{1.5}}{b_1^2} \tag{A1.6-2}$$

Eq. (3.3-3)
Relative inner rotational velocity of trailing string - From Eqs. (1.5-3) and (A1.6-1):

$$\beta_{2r}^{in} = \sqrt{1 - \left(\beta_{2t}^{in}\right)^2} = \sqrt{1 - \frac{2b_1 - 1}{b_1^4}} = \frac{\sqrt{b_1^4 - 2b_1 + 1}}{b_1^2}$$

$$\beta_{2r}^{in} = \frac{\sqrt{b_1^4 - 2b_1 + 1}}{b_1^2} \tag{A1.6-3}$$

Eq. (3.3-4)
Relative outer rotational velocity of trailing string - From Eqs. (1.5-3) and (A1.6-2):

$$\beta_{2r}^{out} = \sqrt{1 - \left(\beta_{2t}^{out}\right)^2} = \sqrt{1 - \frac{(2b_1 - 1)^3}{b_1^4}} = \frac{\sqrt{b_1^4 - (2b_1 - 1)^3}}{b_1^2}$$

$$\beta_{2r}^{out} = \frac{\sqrt{b_1^4 - (2b_1 - 1)^3}}{b_1^2} \tag{A1.6-4}$$

A2 Toryx Parameters as Function of φ_2 (Fig. 1.5)

A2.1 Toryx Radii

Eq. (1.6-14a)
Relative radius of toryx leading string – From Eq. (A1.3-1):

$$\cos s\varphi_2 = \frac{b_1 - 1}{b_1} = 1 - \frac{1}{b_1}; \quad \frac{1}{b_1} = 1 - \cos s\varphi_2$$

$$b_1 = \frac{1}{1 - \cos s\varphi_2} \tag{A2.1-1}$$

Eq. (1.6-14b)
Relative radius of toryx trailing string – From Eqs. (A1.1-2) and (A2.1-1):

$$b_2 = b_1 - 1 = \frac{1}{1 - \cos s\varphi_2} - 1 = \frac{1 - 1 + \cos s\varphi_2}{1 - \cos s\varphi_2} = \frac{\cos s\varphi_2}{1 - \cos s\varphi_2}$$

$$b_2 = \frac{\cos s\varphi_2}{1 - \cos s\varphi_2} \tag{A2.1-2}$$

Eq. (1.6-23)
Relative toryx spherical boundary radius – From Eqs. (A1.1-3) and (A2.1-1):

$$b = 2b_1 - 1 = \frac{2}{1 - \cos s\varphi_2} - 1 = \frac{1 + \cos s\varphi_2}{1 - \cos s\varphi_2}$$

$$b = \frac{1 + \cos s\varphi_2}{1 - \cos s\varphi_2} \tag{A2.1-3}$$

A2.2 Toryx Lengths, Wavelengths & Number of Windings

Eq. (1.6-16a) and (1.6-16b)
Relative length of one winding of toryx leading and trailing strings – From Eqs. (A1.2-1), (A1.2-2) and (A2.1-1):

$$l_1 = l_2 = \frac{1}{1 - \cos s\varphi_2} \tag{A2.2-1}$$

Eq. (1.6-15b)
Relative wavelength of toryx trailing string – From Eqs. (A1.2-3) and (A2.1-1):

$$\eta_2 = \sqrt{2b_1 - 1} = \sqrt{\frac{2}{1 - \cos s\varphi_2} - 1} = 2\sqrt{\frac{1 + \cos s\varphi_2}{1 - \cos s\varphi_2}}$$

$$\eta_2 = \sqrt{\frac{1 + \cos s\varphi_2}{1 - \cos s\varphi_2}} \tag{A2.2-2}$$

Eq. (1.6-17b)
The number of windings of toryx trailing string – From Eqs. (A1.2-4) and (A2.1-1):

$$w_2 = \frac{b_1}{\sqrt{2b_1 - 1}} = \frac{1}{1 - \cos s\varphi_2}\frac{1}{\sqrt{\dfrac{2}{1 - \cos s\varphi_2} - 1}} = \frac{1}{1 - \cos s\varphi_2}\frac{1}{\sqrt{\dfrac{1 + \cos s\varphi_2}{1 - \cos s\varphi_2}}} =$$

$$\frac{1}{\sqrt{\dfrac{(1 - \cos s\varphi_2)^2(1 + \cos s\varphi_2)}{1 - \cos s\varphi_2}}} = \frac{1}{\sqrt{1 - \cos s^2\varphi_2}} = \frac{1}{\sqrt{\sin s^2\varphi_2}} = \frac{1}{\sin s\varphi_2}$$

$$w_2 = \frac{1}{\sin s\varphi_2} \tag{A2.2-3}$$

A2.3 Middle Velocities of Toryx Strings (Fig. 1.5)

Eq. (1.6-18b)
Relative translational velocity of toryx trailing string - From Eqs. (1.5-3) and (1.6-8b):

$$\beta_{2t} = \beta_2 \sin s\varphi_2 = \sin s\varphi_2$$

$$\beta_{2t} = \sin s\varphi_2 \tag{A2.3-1}$$

Eqs. (1.6-19a) and (1.6-20a)
Relative spiral and rotational velocities of toryx leading string – From Eqs. (A1.4-3) and (A2.3-1):

$$\beta_1 = \beta_{1r} = \beta_{2t} = \sin s\varphi_2$$

$$\beta_1 = \beta_{1r} = \sin s\varphi_2 \tag{A2.3-2}$$

Eq. (1.6-19b)
Relative rotational velocity of toryx trailing string – From Eq. (1.5-3):

$$\beta_{2r} = \beta_2 \cos s\varphi_2 = \cos s\varphi_2$$

$$\beta_{2r} = \cos s\varphi_2 \tag{A2.3-3}$$

A2.4 Frequencies of Toryx Strings

Eq. (1.6-21a)
Relative frequency of toryx leading string – From Eqs. (A1.4-2), (A1.5-1), (A2.1-1) and (A2.3-1):

$$\delta_1 = \frac{\sqrt{2b_1-1}}{b_1^2} = \frac{\beta_{2t}}{b_1} = \sin s\varphi_2 (1-\cos s\varphi_2)$$

$$\beta_{2t} = \sin s\varphi_2 (1-\cos s\varphi_2) \tag{A2.4-1}$$

Eq. (1.6-21b)
Relative frequency of toryx trailing string – From Eqs. (A1.5-2) and (A2.1-1):

$$\delta_2 = \frac{1}{b_1} = 1-\cos s\varphi_2$$

$$\delta_2 = 1-\cos s\varphi_2 \tag{A2.4-2}$$

A3 Relationship between Parameters of Toryces of Main Groups

B3.1 Radii of Reality-Polarized Toryces (Fig. 2.3.1)

Eq. (3.2-5a)
Relative radius of leading string of reality-polarized negative toryces $\breve{E}^- \leftrightarrow E^-$ - From Eqs. (2.4-1) and (3.2-1a):

$$\breve{V}_E^- = \frac{1}{V_E^-}; \quad -\frac{\breve{b}_{1E}^- -1}{\breve{b}_{1E}^-} = -\frac{b_{1E}^-}{b_{1E}^- -1}; \quad -1+\frac{1}{\breve{b}_{1E}^-} = -\frac{b_{1E}^-}{b_{1E}^- -1}; \quad \frac{1}{\breve{b}_{1E}^-} = -\frac{b_{1E}^-}{b_{1E}^- -1}+1;$$

$$\frac{1}{\breve{b}_{1E}^-} = \frac{-b_{1E}^- + b_{1E}^- -1}{b_{1E}^- -1} = \frac{-1}{b_{1E}^- -1} = \frac{1}{1-b_{1E}^-}$$

$$\breve{b}_{1E}^- = 1-b_{1E}^- \tag{A3.1-1a}$$

- **Relative radius of leading string of reality-polarized negative toryces** $\breve{A}^- \leftrightarrow A^-$ - Similarly to Eq. (A3.1-1a):

$$\breve{b}_{1A}^- = 1 - b_{1A}^-$$

(A3.1-1b)

Eq. (3.2-6a)

Relative radius of leading string of reality-polarized positive toryces $\breve{E}^+ \leftrightarrow E^+$ - From Eqs. (2.4-1) and (3.2-2a):

$$\breve{V}_E^+ = \frac{1}{V_E^+}; \quad \frac{\breve{b}_{1E}^+ - 1}{\breve{b}_{1E}^+} = \frac{b_{1E}^+}{b_{1E}^+ - 1}; \quad 1 - \frac{1}{\breve{b}_{1E}^+} = \frac{b_{1E}^+}{b_{1E}^+ - 1}; \quad 1 - \frac{b_{1E}^+}{b_{1E}^+ - 1} = \frac{1}{\breve{b}_{1E}^+};$$

$$\frac{b_{1E}^+ - 1 - b_{1E}^+}{b_{1E}^+ - 1} = \frac{1}{\breve{b}_{1E}^+}; \quad \frac{-1}{b_{1E}^+ - 1} = \frac{1}{\breve{b}_{1E}^+}; \quad \breve{b}_{1E}^+ = 1 - b_{1E}^+$$

$$\breve{b}_{1E}^+ = 1 - b_{1E}^+$$

(A3.1-2a)

- **Relative radius of leading string of reality-polarized positive toryces** $\breve{A}^+ \leftrightarrow A^+$ - Similarly to Eq. (A3.1-2a):

$$\breve{b}_{1A}^+ = 1 - b_{1A}^+$$

(A3.1-2b)

Eq. (3.2-5b)

Relative radius of spherical boundary of reality-polarized negative toryces $\breve{E}^- \leftrightarrow E^-$ - From Eqs. (1.6-13) and (A3.1-1a):

$$\breve{b}_E^- = 2\breve{b}_{1E}^- - 1 = 2(1 - b_{1E}^-) - 1 = 2 - 2b_{1E}^- - 1 = 1 - 2b_{1E}^- =$$

$$1 - 2\frac{b_E^- + 1}{2} = 1 - b_E^- - 1 = -b_E^-$$

$$\breve{b}_E^- = -b_E^-$$

(A3.1-3a)

- **Relative radius of spherical boundary of reality-polarized negative toryces** $\breve{A}^- \leftrightarrow A^-$ - Similarly to Eq. (A3.1-3a):

$$\breve{b}_A^- = -b_A^-$$

(A3.1-3b)

Eq. (3.2-6b)

Relative radius of spherical boundary of reality-polarized positive toryces $\breve{E}^+ \leftrightarrow E^+$ - From Eqs. (1.6-13) and (A3.1-2a):

$$\breve{b}_E^+ = 2\breve{b}_{1E}^+ - 1 = 2(1 - b_{1E}^+) - 1 = 2 - 2b_{1E}^+ - 1 = 1 - 2b_{1E}^+ =$$

$$1 - 2\frac{b_E^+ + 1}{2} = 1 - b_E^+ - 1 = -b_E^+$$

$$\breve{b}_E^+ = -b_E^+ \tag{A3.1-4a}$$

- **Relative radius of spherical boundary of reality-polarized positive toryces** $\breve{A}^+ \leftrightarrow A^+$ - Similarly to Eq. (A3.1-4a):

$$\breve{b}_A^+ = -b_A^+ \tag{A3.1-4b}$$

A3.2 Radii of Vorticity-Polarized Toryces (Fig. 2.3.1)

Eq. (3.2-7a)
Relative radius of leading string of vorticity-polarized real toryces $A^+ \leftrightarrow A^-$ - From Eqs. (1.6-13), (2.4-1) and (3.2-3a):

$$V_A^+ = -V_A^-; \quad -\frac{b_{1A}^+ - 1}{b_{1A}^+} = \frac{b_{1A}^- - 1}{b_{1A}^-}; \quad -1 + \frac{1}{b_{1A}^+} = 1 - \frac{1}{b_{1A}^-}; \quad \frac{1}{b_{1A}^+} = 2 - \frac{1}{b_{1A}^-};$$

$$\frac{1}{b_{1A}^+} = \frac{2b_{1A}^- - 1}{b_{1A}^-}; \quad b_{1A}^+ = \frac{b_{1A}^-}{2b_{1A}^- - 1}$$

$$b_{1A}^+ = \frac{b_{1A}^-}{2b_{1A}^- - 1} \tag{A3.2-1a}$$

- **Relative radius of leading string of vorticity-polarized real toryces** $E^+ \leftrightarrow E^-$ - Similarly to Eq. (A3.2-1a):

$$b_{1E}^+ = \frac{b_{1E}^-}{2b_{1E}^- - 1} \tag{A3.2-1b}$$

Eq. (3.2-8a)
Relative radius of leading string of vorticity-polarized imaginary toryces $\breve{A}^+ \leftrightarrow \breve{A}^-$ - From Eqs. (2.4-1), (3.2-4a):

$$\breve{V}_A^+ = -\breve{V}_A^-; \quad -\frac{\breve{b}_{1A}^+ - 1}{\breve{b}_{1A}^+} = \frac{\breve{b}_{1A}^- - 1}{\breve{b}_{1A}^-}; \quad -1 + \frac{1}{\breve{b}_{1A}^+} = 1 - \frac{1}{\breve{b}_{1A}^-}; \quad \frac{1}{\breve{b}_{1A}^+} = 2 - \frac{1}{\breve{b}_{1A}^-};$$

$$\frac{1}{\breve{b}_{1A}^+} = \frac{2\breve{b}_{1A}^- - 1}{\breve{b}_{1A}^-}; \quad \breve{b}_{1A}^+ = \frac{\breve{b}_{1A}^-}{2\breve{b}_{1A}^- - 1}$$

$$\breve{b}_{1A}^+ = \frac{\breve{b}_{1A}^-}{2\breve{b}_{1A}^- - 1} \tag{A3.2-1a}$$

- **Relative radius of leading string of vorticity-polarized imaginary toryces** $\breve{E}^+ \leftrightarrow \breve{E}^-$ - Similarly to Eq. (A3.2-1a):

$$\breve{b}_{1E}^{+} = \frac{\breve{b}_{1E}^{-}}{2\breve{b}_{1E}^{-} - 1} \qquad \text{(A3.2-1b)}$$

Eq. (3.2-7b)

Relative radius of spherical boundary of vorticity-polarized real toryces $A^{+} \leftrightarrow A^{-}$ - From Eqs. (1.6-13) and (A3.2-1a):

$$b_{A}^{+} = 2b_{1A}^{+} - 1 = \frac{2b_{1A}^{-}}{2b_{1A}^{-} - 1} - 1 = \frac{2b_{1A}^{-} - 2b_{1A}^{-} + 1}{2b_{1A}^{-} - 1} = \frac{1}{2b_{1A}^{-} - 1} = \frac{1}{b_{A}^{-}}$$

$$b_{A}^{+} = \frac{1}{b_{A}^{-}} \qquad \text{(A3.2-2a)}$$

- **Relative radius of spherical boundary of vorticity-polarized real toryces** $E^{+} \leftrightarrow E^{-}$ -
  Similarly to Eq. (A3.2-2a):

$$b_{E}^{+} = \frac{1}{b_{E}^{-}} \qquad \text{(A3.2-2b)}$$

Eq. (3.2-8b)

Relative radius of spherical boundary of vorticity-polarized imaginary toryces $\breve{A}^{+} \leftrightarrow \breve{A}^{-}$ -
From Eqs. (1.6-13) and (A3.2-1a):

$$\breve{b}_{A}^{+} = 2\breve{b}_{1A}^{+} - 1 = \frac{2\breve{b}_{1A}^{-}}{2\breve{b}_{1A}^{-} - 1} - 1 = \frac{2\breve{b}_{1A}^{-} - 2\breve{b}_{1A}^{-} + 1}{2\breve{b}_{1A}^{-} - 1} = \frac{1}{2\breve{b}_{1A}^{-} - 1} = \frac{1}{\breve{b}_{A}^{-}}$$

$$\breve{b}_{A}^{+} = \frac{1}{\breve{b}_{A}^{-}} \qquad \text{(A3.2-3a)}$$

- **Relative radius of spherical boundary of vorticity-polarized real toryces** $\breve{E}^{+} \leftrightarrow \breve{E}^{-}$ -
  Similarly to Eq. (A3.2-3a):

$$\breve{b}_{E}^{+} = \frac{1}{\breve{b}_{E}^{-}} \qquad \text{(A3.2-3b)}$$

A4 Relationship between Parameters of Toryces of Subgroups

Eq. (3.2-14a)

Relative radius of leading string of real negative toryces $A^{-} \leftrightarrow E^{-}$ - From Eqs. (2.4-1) and (3.2-10):

$$V_A^- + V_E^- = -1; \quad \frac{b_{1A}^- - 1}{b_{1A}^-} + \frac{b_{1E}^- - 1}{b_{1E}^-} = 1; \quad 1 - \frac{1}{b_{1A}^-} + 1 - \frac{1}{b_{1E}^-} = 1; \quad \frac{1}{b_{1A}^-} = 1 - \frac{1}{b_{1E}^-};$$

$$\frac{1}{b_{1A}^-} = \frac{b_{1E}^- - 1}{b_{1E}^-}; \quad b_{1A}^- = \frac{b_{1E}^-}{b_{1E}^- - 1}$$

$$b_{1A}^- = \frac{b_{1E}^-}{b_{1E}^- - 1} \tag{A4.1-1}$$

Eq. (3.2-15a)

Relative radius of leading string of real positive toryces $A^+ \leftrightarrow E^+$ - From Eqs. (2.4-1) and (3.2-11):

$$V_A^+ + V_E^+ = 1; \quad -\frac{b_{1A}^+ - 1}{b_{1A}^+} - \frac{b_{1E}^+ - 1}{b_{1E}^+} = 1; \quad -1 + \frac{1}{b_{1A}^+} - 1 + \frac{1}{b_{1E}^+} = 1; \quad \frac{1}{b_{1A}^+} = 3 - \frac{1}{b_{1E}^+};$$

$$\frac{1}{b_{1A}^+} = \frac{3b_{1E}^+ - 1}{b_{1E}^+}; \quad b_{1A}^+ = \frac{b_{1E}^+}{3b_{1E}^+ - 1}$$

$$b_{1A}^+ = \frac{b_{1E}^+}{3b_{1E}^+ - 1} \tag{A4.1-2}$$

Eq. (3.2-14b)

Relative radius of spherical boundary of real negative toryces $A^- \leftrightarrow E^-$ - From Eqs. (1.6-13) and (A4.1-1):

$$b_A^- = 2b_{1A}^- - 1 = \frac{2b_{1E}^-}{b_{1E}^- - 1} - 1 = \frac{2(b_E^- + 1)}{2} \frac{1}{\frac{b_E^- + 1}{2} - 1} - 1 = \frac{2(b_E^- + 1)}{b_E^- + 1 - 2} - 1 = \frac{2(b_E^- + 1)}{b_E^- - 1} - 1;$$

$$\frac{2b_E^- + 2}{b_E^- - 1} - 1 = \frac{2b_E^- + 2 - b_E^- + 1}{b_E^- - 1} = \frac{b_E^- + 3}{b_E^- - 1}$$

$$b_A^- = \frac{b_E^- + 3}{b_E^- - 1} \tag{A4.1-3}$$

Eq. (3.2-15b)

Relative radius of spherical boundary of real positive toryces $A^+ \leftrightarrow E^+$ - From Eqs. (1.6-13) and (A4.1-2):

$$b_A^+ = 2b_{1A}^+ - 1 = \frac{2b_{1E}^+}{3b_{1E}^+ - 1} - 1 = \frac{2(b_E^+ + 1)}{2} \frac{1}{\frac{3(b_E^+ + 1)}{2} - 1} - 1 = \frac{2(b_E^+ + 1)}{3b_E^+ + 3 - 2} - 1 = \frac{2b_E^+ + 2}{3b_E^+ + 1} - 1;$$

$$\frac{2b_E^+ + 2}{3b_E^+ + 1} - 1 = \frac{2b_E^+ + 2 - 3b_E^+ - 1}{3b_E^+ + 1} = \frac{-b_E^+ + 1}{3b_E^+ + 1}$$

$$b_A^+ = \frac{1 - b_E^+}{1 + 3b_E^+} \tag{A4.1-4}$$

A5 Excitation Quantum States of Toryces

Eq. (6.2-1) - From Eq. (6.1-1):

$$b_{1E}^- = z \tag{A5-1}$$

Eq. (6.2-2) - From Eqs. (A5-1) and (1.6-13):

$$b_E^- = 2b_E^- - 1 = 2z - 1$$
$$b_E^- = 2z - 1 \tag{A5-2}$$

Eq. (6.2-3) - From Eqs. (A3.1-1a) and (A5-1):

$$\breve{b}_{1E}^- = 1 - b_{1E}^- = 1 - z$$
$$b_{1E}^- = 1 - z \tag{A5-3}$$

Eq. (6.2-4) - From Eqs. (A3.1-3a) and (A5-2):

$$\breve{b}_E^- = -b_E^- = 1 - 2z$$
$$\breve{b}_E^- = 1 - 2z \tag{A5-4}$$

Eq. (6.2-5) - From Eqs. (A3.2-1b) and (A5-1):

$$b_{1E}^+ = \frac{b_{1E}^-}{2b_{1E}^- - 1} = \frac{z}{2z - 1}$$

$$b_{1E}^{+} = \frac{z}{2z - 1} \tag{A5-5}$$

Eq. (6.2-6) - From Eqs. (A3.2-2b) and (A5-2):

$$b_{E}^{+} = \frac{1}{b_{E}^{-}} = \frac{1}{2z - 1}$$

$$b_{E}^{-} = \frac{1}{2z - 1} \tag{A5-6}$$

Eq. (6.2-7) - From Eqs. (A3.1-2a) and (A5-5):

$$\breve{b}_{1E}^{+} = 1 - b_{1E}^{+} = 1 - \frac{z}{2z - 1} = \frac{2z - 1 - z}{2z - 1} = \frac{z - 1}{2z - 1} = \frac{1 - z}{1 - 2z}$$

$$b_{1E}^{+} = \frac{1 - z}{1 - 2z} \tag{A5-7}$$

Eq. (6.2-8) - From Eqs. (A3.2-3b) and (A5-4):

$$\breve{b}_{E}^{+} = \frac{1}{\breve{b}_{E}^{-}} = \frac{1}{1 - 2z}$$

$$\breve{b}_{E}^{+} = \frac{1}{1 - 2z} \tag{A5-8}$$

Eq. (6.2-9) - From Eqs. (A4.1-1) and (A5-1):

$$b_{1A}^{-} = \frac{b_{1E}^{-}}{b_{1E}^{-} - 1} = \frac{z}{z - 1}$$

$$b_{1A}^{-} = \frac{z}{z - 1} \tag{A5-9}$$

Eq. (6.2-10) - From Eqs. (1.6-13) and (A5-9):

$$b_{A}^{-} = 2b_{1A}^{-} - 1 = \frac{2z}{z - 1} - 1 = \frac{2z - z + 1}{z - 1} = \frac{z + 1}{z - 1}$$

$$b_A^- = \frac{z+1}{z-1} \qquad (A5\text{-}10)$$

Eq. (6.2-11) - From Eqs. (A3.2-1a) and (A5-9):

$$b_{1A}^+ = \frac{b_{1A}^-}{2b_{1A}^- - 1} = \frac{z}{z-1} \frac{1}{\dfrac{2z}{z-1} - 1} = \frac{z}{2z - z + 1} = \frac{z}{z+1}$$

$$b_{1A}^+ = \frac{z}{z+1} \qquad (A5\text{-}11)$$

Eq. (6.2-12) - From Eqs. (A3.2-2a) and (A5-10):

$$b_A^+ = \frac{1}{b_A^-} = \frac{z-1}{z+1}$$

$$b_A^+ = \frac{z-1}{z+1} \qquad (A5\text{-}12)$$

Eq. (6.2-13) - From Eqs. (A3.1-1b) and (A5-9):

$$\breve{b}_{1A}^- = 1 - b_{1A}^- = 1 - \frac{z}{z-1} = \frac{z-1-z}{z-1} = \frac{-1}{z-1} = \frac{1}{1-z}$$

$$\breve{b}_{1A}^- = \frac{1}{1-z} \qquad (A5\text{-}13)$$

Eq. (6.2-14) - From Eqs. (1.6-13) and (A5-13):

$$\breve{b}_A^- = 2\breve{b}_{1A}^- - 1 = \frac{2}{1-z} - 1 = \frac{2-1+z}{1-z} = \frac{1+z}{1-z}$$

$$\breve{b}_A^- = \frac{1+z}{1-z} \qquad (A5\text{-}14)$$

Eq. (6.2-15) - From Eqs. (A3.2-1a) and (A5-13):

$$\breve{b}_{1A}^+ = \frac{\breve{b}_{1A}^-}{2\breve{b}_{1A}^- - 1} = \frac{1}{1-z} \frac{1}{\dfrac{2}{1-z} - 1} = \frac{1}{2-1+z} = \frac{1}{1+z}$$

$$\breve{b}_{1A}^{+} = \frac{1}{1+z}$$

(A5-15)

Eq. (6.2-16) - From Eqs. (A3.2-3a) and (A5-14):

$$\breve{b}_{A}^{+} = \frac{1}{\breve{b}_{A}^{-}} = \frac{1-z}{1+z}$$

$$b_{A}^{+} = \frac{1-z}{1+z}$$

(A5-16)

A6 Toryx Reality Ratio

Eq. (2.4-1)
Vorticity of real toryces – By definition:

$$V = -\frac{b_1 - 1}{b_1}$$

(A6-1)

- **Vorticity of imaginary toryces** – From Eqs. (A6-1), (3.2-5a) and (3.2-6a):

$$\breve{V} = -\frac{\breve{b}_1 - 1}{\breve{b}_1} = -\frac{1 - b_1 - 1}{1 - b_1} = -\frac{b_1}{b_1 - 1}$$

$$\breve{V} = -\frac{b_1}{b_1 - 1}$$

(A6-2)

Eq. (7.4-3) – **Toryx reality ratio** – From Eqs. (A6-1) and (A6-2):

$$T = \frac{\breve{V}}{V} = \frac{N}{\breve{N}} = \left(-\frac{b_1}{b_1-1}\right)\left(-\frac{b_1}{b_1-1}\right) = \left(\frac{b_1}{b_1-1}\right)^2$$

$$T = \frac{\breve{V}}{V} = \frac{N}{\breve{N}} = \left(\frac{b_1}{b_1-1}\right)^2$$

(A6-3)

A7 Instant Velocities of Trailing String (Fig. 3.3.4)

- Increment of relative distance Δb along axis X within trailing string:

$$\Delta b = \frac{2(b_1 - 1)}{m} \tag{A7-1}$$

- X-coordinate x_m of middle point m of increment z_m:

$$x_m = x_{m-1} + \Delta b \tag{A7-2}$$

- X-coordinate x_{m1} of entry point m_1 of increment z_m:

$$x_{m1} = x_m - \frac{\Delta b}{2} \tag{A7-3}$$

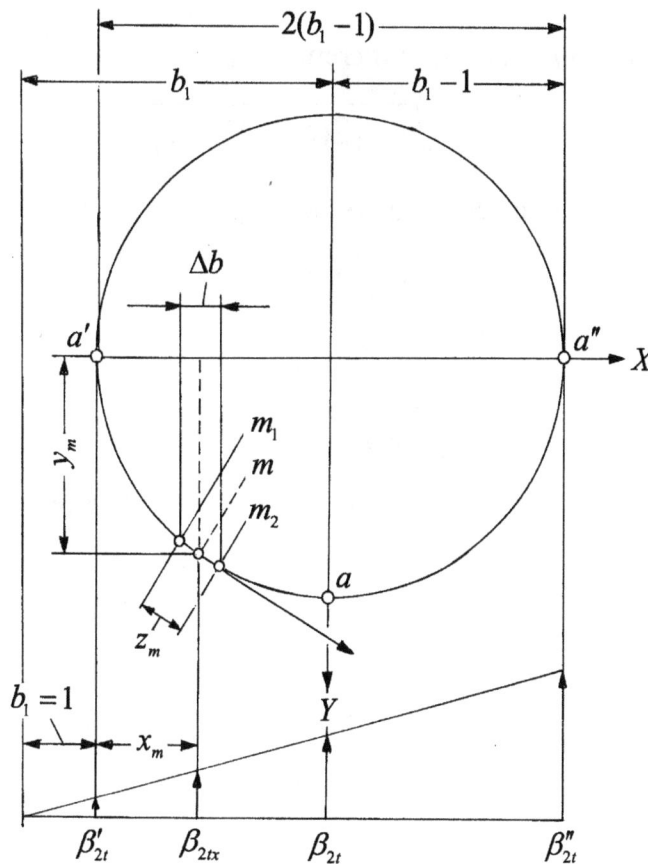

Figure 3.3.4. Derivation of instantaneous velocities of trailing string.

- Increment of relative distance Δb along axis X within trailing string:

$$\Delta b = \frac{2(b_1 - 1)}{m} \qquad \text{(A7-1)}$$

- X-coordinate x_m of middle point m of increment z_m:

$$x_m = x_{m-1} + \Delta b \qquad \text{(A7-2)}$$

- X-coordinate x_{m1} of entry point m_1 of increment z_m:

$$x_{m1} = x_m - \frac{\Delta b}{2} \qquad \text{(A7-3)}$$

- X-coordinate x_{m2} of exit point m_2 of increment z_m:

$$x_{m2} = x_m + \frac{\Delta b}{2} \qquad \text{(A7-4)}$$

- Y-coordinate y_{m1} of entry point m_1 of increment z_m:

$$y_{m1} = \sqrt{x_{m1}[2(b_1 - 1) - x_{m1}]} \qquad \text{(A7-5)}$$

- Y-coordinate y_{m2} of exit point m_2 of increment z_m:

$$y_{m2} = \sqrt{x_{m1}[2(b_1 - 1) - x_{m2}]} \qquad \text{(A7-6)}$$

- Increment z_m:

$$z_m = \sqrt{(\Delta b)^2 + (y_{m2} - y_{m1})^2} \qquad \text{(A7-7)}$$

- Relative translational velocity β_{2tx} of point m - From Eq. (1.6-6b):
-

$$\beta_{2tx} = \beta_{2t} \frac{x_m + 1}{b_1} = (x_m + 1)\frac{\sqrt{2b_1 - 1}}{b_1^2} \qquad \text{(A7-8)}$$

- Relative rotational velocity β_{2rx} of point m – From Eq. (1.5-3):
-

$$\beta_{2rx} = \sqrt{1 - \beta_{2tx}^2} \qquad \text{(A7-9)}$$

- Increment of time Δt_m corresponding to increment z_m – From Eq. (1.4-18):

$$\Delta t_m = \frac{z_m r_0}{\beta_{2rx} c} = \frac{z_m}{\beta_{2rx}} \frac{2\pi r_0}{2\pi c} = \frac{z_m}{\beta_{2rx}} \frac{1}{2\pi f_0}$$

$$\Delta t_m = \frac{z_m}{\beta_{2rx}} \frac{1}{2\pi f_0} \tag{A7-10}$$

- Relative increment of time $\Delta t_m / T_2$ corresponding to z_m:

$$\frac{\Delta t_m}{T_2} = \frac{z_m}{\beta_{2rx}} \frac{1}{2\pi f_0} \frac{f_0}{b_1} = \frac{z_m}{\beta_{2rx}} \frac{1}{2\pi b_1}$$

$$\frac{\Delta t_m}{T_2} = \frac{z_m}{\beta_{2rx}} \frac{1}{2\pi b_1} \tag{A7-11}$$

- Maximum increment of relative translational velocity $\Delta\beta_{2tmax}$:

$$\Delta\beta_{2t\,max} = \beta'_{2t} - \beta_{2t} = \frac{\sqrt{2b_1 - 1}}{b_1^2} - \frac{\sqrt{2b_1 - 1}}{b_1} = -\frac{(b_1 - 1)\sqrt{2b_1 - 1}}{b_1^2}$$

$$\Delta\beta_{2t\,max} = \beta'_{2t} - \beta_{2t} = -\frac{(b_1 - 1)\sqrt{2b_1 - 1}}{b_1^2} \tag{A7-12}$$

- Instantaneous increments of relative translational velocity $\Delta\beta_{2t}$:

$$\Delta\beta_{2t} = \Delta\beta_{2t\,max} \cos\left(\frac{2\pi f_0}{b_1} t\right) = -\frac{(b_1 - 1)\sqrt{2b_1 - 1}}{b_1^2} \cos\left(\frac{2\pi f_0}{b_1} t\right)$$

$$\Delta\beta_{2t} = -\frac{(b_1 - 1)\sqrt{2b_1 - 1}}{b_1^2} \cos\left(\frac{2\pi f_0}{b_1} t\right) \tag{A7-13}$$

Notes

Notes

Notes

Appendix B

DERIVATION OF HELYX EQUATIONS

B1 Helyx Parameters as Functions of $\tilde{l}_2$

Table 12.4. Helyx relative spacetime parameters.

Toryx relative parameters	Equations
Radius of helyx cylindrical boundary	$\tilde{b} = \tilde{r} / \tilde{r}_0$ (12.4-1)
Radius of helyx eye	$\tilde{b}_0 = 1$ (12.4-2)
Radius of helyx leading string	$\tilde{b}_1 = \tilde{r}_1 / \tilde{r}_0$ (12.4-3)
Radius of helyx trailing string	$\tilde{b}_2 = \tilde{r}_2 / \tilde{r}_0$ (12.4-4)
Length of helyx leading string	$\tilde{l}_1 = \tilde{L}_1 / 2\pi\tilde{r}_0$ (12.4-5)
Length of helyx trailing string	$\tilde{l}_2 = \tilde{L}_2 / 2\pi\tilde{r}_0$ (12.4-6)
Period of helyx leading string	$\tilde{t}_1 = \tilde{T}_1 / \tilde{f}_0$ (12.4-7)
Period of helyx trailing string	$\tilde{t}_2 = \tilde{T}_2 / \tilde{f}_0$ (12.4-8)
Spiral velocity of helyx leading string	$\tilde{\beta}_1 = \tilde{V}_1 / c$ (12.4-9)
Translational velocity of helyx leading string	$\tilde{\beta}_{1t} = \tilde{V}_{1t} / c$ (12.4-10)
Rotational velocity of helyx leading string	$\tilde{\beta}_{1r} = \tilde{V}_{1r} / c$ (12.4-11)
Spiral velocity of helyx trailing string	$\tilde{\beta}_2 = \tilde{V}_2 / c$ (12.4-12)
Translational velocity of helyx trailing string	$\tilde{\beta}_{2t} = \tilde{V}_{2t} / c$ (12.4-13)
Rotational velocity of helyx trailing string	$\tilde{\beta}_{2r} = \tilde{V}_{2r} / c$ (12.4-14)
Frequency of helyx leading string	$\tilde{\delta}_1 = \tilde{f}_1 / \tilde{f}_0$ (12.4-15)
Frequency of helyx trailing string	$\tilde{\delta}_2 = \tilde{f}_2 / \tilde{f}_0$ (12.4-16)
Wavelength of helyx leading string	$\tilde{\eta}_1 = \tilde{\lambda}_1 / 2\pi\tilde{r}_0$ (12.4-17)
Wavelength of helyx trailing string	$\tilde{\eta}_2 = \tilde{\lambda}_2 / 2\pi\tilde{r}_0$ (12.4-18)

Eq. (12.6-3a)
Relative wavelength of trailing string – From Eq. (12.5-2):

$$\tilde{\eta}_2 = \tilde{l}_2 - 1 \tag{B1-1}$$

Eq. (12.6-4a)
Relative radius of trailing string – From Eq. (B1-1) and Fig. 12.5:

$$\tilde{b}_2 = \sqrt{\tilde{l}_2^{\,2} - \tilde{\eta}_2^{\,2}} = \sqrt{\tilde{l}_2^{\,2} - (\tilde{l}_2 - 1)^2} = \sqrt{2\tilde{l}_2 - 1}$$

$$\tilde{b}_2 = \sqrt{2\tilde{l}_2 - 1} \tag{B1-2}$$

Eq. (12.6-9a)
Frequency of trailing string – From Eqs. (12.3-3), (12.3-4), (12.4-6) and (12.4-16):

$$\tilde{f}_2 = \tilde{\delta}_2 \tilde{f}_0 = \frac{\tilde{V}_2}{\tilde{L}_2} = \frac{c}{\tilde{l}_2\, 2\pi \tilde{r}_0} = \frac{\tilde{f}_0}{\tilde{l}_2} \tag{B1-3}$$

Relative frequency of trailing string – From Eq. (B1-3):

$$\tilde{\delta}_2 = \frac{1}{\tilde{l}_2} \tag{B1-4}$$

Eq. (12.6-1a)
Cosine of apex angle of trailing string – From Eq. (B1-1) and Fig.12.5:

$$\cos s\tilde{\varphi}_2 = \frac{\tilde{l}_2 - 1}{\tilde{l}_2} \tag{B1-5}$$

Eq. (12.6-6a)
Relative translational velocity of trailing string – From Eqs. (12.5-3), (B1-5) and Fig.12.5:

$$\tilde{\beta}_{2t} = \tilde{\beta}_2 \cos s\tilde{\varphi}_2 = \frac{\tilde{l}_2 - 1}{\tilde{l}_2}$$

$$\tilde{\beta}_{2t} = \frac{\tilde{l}_2 - 1}{\tilde{l}_2} \tag{B1-6}$$

Eq. (12.6-7a)
Relative rotational velocity of trailing string – From Eqs. (12.5-3), (B1-6) and Fig.12.5:

$$\widetilde{\beta}_{2r} = \widetilde{\beta}_2 \sin s\widetilde{\varphi}_2 = \sqrt{1 - \frac{(\widetilde{l}_2 - 1)^2}{\widetilde{l}_2^{\,2}}} = \frac{\sqrt{2\widetilde{l}_2 - 1}}{\widetilde{l}_2}$$

$$\widetilde{\beta}_{2r} = \frac{\sqrt{2\widetilde{l}_2 - 1}}{\widetilde{l}_2} \tag{B1-7}$$

B2 Helyx Parameters as Functions of $\widetilde{\varphi}_2$

Eq. (12.6-2b)
Relative length of trailing string – From Eq. (B1-5):

$$\cos s\widetilde{\varphi}_2 = \frac{\widetilde{l}_2 - 1}{\widetilde{l}_2} = 1 - \frac{1}{\widetilde{l}_2}; \quad \frac{1}{\widetilde{l}_2} = 1 - \cos s\widetilde{\varphi}_2; \quad \widetilde{l}_2 = \frac{1}{1 - \cos s\widetilde{\varphi}_2}$$

$$\widetilde{l}_2 = \frac{1}{1 - \cos s\widetilde{\varphi}_2} \tag{B2-1}$$

Eq. (12.6-3b)
Relative wavelength of trailing string – From Eqs. (B1-1) and (B2-1):

$$\widetilde{\eta}_2 = \widetilde{l}_2 - 1 = \frac{1}{1 - \cos s\widetilde{\varphi}_2} - 1 = \frac{1 - 1 + \cos s\widetilde{\varphi}_2}{1 - \cos s\widetilde{\varphi}_2} = \frac{\cos s\widetilde{\varphi}_2}{1 - \cos s\widetilde{\varphi}_2}$$

$$\widetilde{\eta}_2 = \frac{\cos s\widetilde{\varphi}_2}{1 - \cos s\widetilde{\varphi}_2} \tag{B2-2}$$

Eq. (12.6-4b)
Relative radius of trailing string – From Eqs. (B1-2) and (B2-1):

$$\widetilde{b}_2 = \sqrt{2\widetilde{l}_2 - 1} = \sqrt{\frac{2}{1 - \cos s\widetilde{\varphi}_2} - 1} = \sqrt{\frac{2 - 1 + \cos s\widetilde{\varphi}_2}{1 - \cos s\widetilde{\varphi}_2}} = \sqrt{\frac{1 + \cos s\widetilde{\varphi}_2}{1 - \cos s\widetilde{\varphi}_2}}$$

$$\widetilde{b}_2 = \sqrt{\frac{1 + \cos s\widetilde{\varphi}_2}{1 - \cos s\widetilde{\varphi}_2}} \tag{B2-3}$$

Eq. (12.6-9b)
Relative frequency of trailing string – From Eqs. (B1-4) and (B2-1):

$$\widetilde{\delta}_2 = \frac{1}{\widetilde{l}_2} = 1 - \cos s\widetilde{\varphi}_2$$

$$\widetilde{\delta}_2 = 1 - \cos s\widetilde{\varphi}_2 \tag{B2-4}$$

Eq. (12.6-6b)
Relative translational velocity of trailing string – From Eq. (12.5-3) and Fig.12.5:

$$\widetilde{\beta}_{2t} = \cos s\widetilde{\varphi}_2 \tag{B2-5}$$

Eq. (12.6-7b)
Relative rotational velocity of trailing string – From Eq. (12.5-3) and Fig.12.5:

$$\widetilde{\beta}_{2r} = \sin s\widetilde{\varphi}_2 \tag{B2-6}$$

<u>Notes</u>

Notes

PHYSICAL & SPACETIME CONSTANTS

Physical Constants

Names		Values
Elementary charge	e	$1.602\,176\,565 \times 10^{-19}$ C
Electric constant	ε_0	$8.854\,187\,817 \times 10^{-12}$ C^2/N/m^2
Electron mass	m_e	$9.109\,382\,91 \times 10^{-31}$ kg
Proton relative mass	m_p / m_e	$1836.152\,671\,95$
Neutron relative mass	m_n / m_e	$1838.683\,660\,08$
Muon relative mass	m_μ / m_e	$206.768\,284\,26$
Tau relative mass	m_τ / m_e	$3477.151\,011\,54$
Newtonian constant of gravitation	G	$6.674\,280 \times 10^{-11}$ m^3/kg/s^2
Planck constant	h	$6.626\,069\,605 \times 10^{-34}$ J s
Bohr magneton	μ_B	$9.274\,009\,68 \times 10^{-24}$ J/T
Muon magneton	μ_μ	$4.485\,218\,58 \times 10^{-26}$ J/T
Tau magneton	μ_τ	$2.666\,876\,44 \times 10^{-24}$ J/T
Electron magnetic moment to Bohr magneton ratio	μ_e / μ_B	$-1.001\,159\,652\,180\,76$
Muon magnetic moment to muon magneton ratio	μ_μ / μ_μ	$-1.001\,165\,923$
Tau magnetic moment to tau magneton ratio	μ_τ / μ_τ	$-0.987\,629\,486$
Nuclear magneton	μ_N	$5.050\,783\,54 \times 10^{-27}$ J/T
Proton magnetic moment to nuclear magneton ratio	μ_p / μ_N	$2.792\,847\,356$
Neutron magnetic moment to nuclear magneton ratio	μ_n / μ_N	$-1.913\,042\,72$

Spacetime Constants

Names		Values
Speed of light in vacuum	c	$2.997\ 924\ 58 \times 10^{08}$ m/s
Classical electron radius	r_e	$2.817\ 940\ 327 \times 10^{-15}$ m
Micro-toryx eye radius	r_0	$1.408\ 970\ 164 \times 10^{-15}$ m
Toryx basic frequency	f_0	$3.386\ 406\ 102 \times 10^{22}$ s^{-1}
Toryx quantization constant	Λ	137
Rydberg constant	R_x	$1.097\ 373\ 157 \times 10^{07}$ m^{-1}
Planck length	l_p	$1.616252(81) \times 10^{-35}$ m

References

List of Publications Consulted

A

Aczel, A.D., *The Mystery of the Aleph*, Washington Square Press, Published by Pocket Books, New York, London, Toronto, Sydney, 2001.

Aczel, A.D., *Entanglement*, A Plume Book, Penguin Group, New York, 2003.

Adair, R.K., *The Great Design – Particles, Fields and Creation*, Oxford University Press, New York, Oxford, 1987.

Agassi, J., *Faraday as a Natural Philosopher*, The University of Chicago Press, Chicago, IL, 1971.

Aiton, E.J., *The Vortex Theory of Planetary Motions*, American Elsevier, Inc., New York, 1972.

Aiton, E.J., *Leibniz - A Biography*, Adam Hilger Ltd, Bristol and Boston, 1985.

Akimov, A.E. and Shipov, G.I., "Torsion Fields and Their Experimental Manifestations," *Proc. of the Int'l Conference on New Ideas in Natural Sciences*, St. Petersburg, Russia, June 1996.

Akimov, A.E. and Tarasenko, V.Y., "Models of Polarized States of the Physical Vacuum and Torsion Fields," *Fizika*, No. 3, March 1992.

Albert, D.Z., "Bohm's Alternative to Quantum Mechanics," *Scientific American*, May 1994.

Albert, D.Z., *Quantum Mechanics and Experience*, Harvard University Press, Cambridge, Massachusetts, London, England, 1992,

Alexandersson, O., *Living Water - Victor Schauberger and the Secrets of Natural Energy*, Gateway Books, Bath, UK, 1996.

Alfven, H., *Worlds - Antiworlds: Antimatter in Cosmology*, W.H. Freeman and Company, San Francisco, 1966.

Allen, H.S., "The Case for a Ring Electron, *Proc. Phys. Soc. London*, vol. 31, pp 49-68 (1919).

Allen, R.E., *Greek Philosophy: Thales to Aristotle*, The Free Press, New York, 1966.

Allexander, A., *Infinitesimal*, Scientific American, Farrar, Straus and Giroux, New York, 2014.

Andrade, E.N. da C., *Rutherford and the Nature of the Atom*, Doubleday & Company, Inc., New York, 1964.

Andrulis, E.D., "Theory of the Origin, Evolution, and Nature of Life," (2012), *Life* **2012**, *2*, 1-30.

Arp, H., *Seeing Red - Redshifts, Cosmology and Academic Science*, Apeiron, Montreal, Canada, 1998.

Ash, D. and Hewitt, P., *The Vortex - Key to Future Science*, Gateway Books, Bath, England, 1991.

B

Babbitt, E.D., *The Principles of Light and Color*, Babbitt & Co., Kessinger Legacy Reprints, 1878.

Baggott, J., *The Meaning of Quantum Theory*, Oxford University Press, Oxford, UK, 1992.

Baker, J., *50 Physics Ideas You Really Need to Know*, Quercus, London, 2007.

Barrow, J.D. and Silk, J., *The Left Hand of Creation*, Oxford University Press, New York, 1983.

Barrow, J.D., *The Constants of Nature – The Numbers that Encode the Deepest Secrets of the Universe*, Vintage Books A Division of Random House, Inc., New York, 2002.

Barrow, J.D., *The Infinite Book – A Short Guide to the Boundless, Timeless and Endless*, Pantheon Books, New York, 2005.

Barrow, J.D., *New Theories of Everything – The Quest for Ultimate Explanation*, Oxford University Press Inc., New York, 2007.

Bartusiak, M., "Loops of Space," *Discover*, April 1993.

Bartusiak, M., "Gravity Wave Sky," *Discover*, July 1993.

Bastrukov, S.I. at al, "Spiral Magneto-Electron Waves in Interstellar Gas," Journal of Experimental and Theoretical Physics, Volume 93, pp 671-676, October 2001.

Beckmann, P., *A History of PI*, Dorset Press, New York, 1989.

Beiser, G., *The Story of Gravity - An Historical Approach to the Study of the Force That Holds the Universe Together*, E.P. Dutton & Co., Inc., New York, 1968.

Bekenstein, J.D., "Information in the Holographic Universe," *Scientific American*, pp. 59-65, August 2003.

Bentov, I., *Stalking the Wild Pendulum – On the Mechanics of Consciousness*, Density Book, Rochester, New York, 1977.

Bergman, D.L. and Wesley, J.P., "Spinning Charge Ring Model of Electron Yielding Anomalous Magnetic Moment," *Galilean Electrodynamics*, Vol. 1, No. 5, Sept./Oct. 1990.

Berke, J.P., Author, Editor, *Nanotubes and Nanowires (Selected Topics in Electronics and Systems*, World Scientific Publishing Company, Singapore, 2007.

Bernauer, J.C. and Paul, R., "The Proton Radius Problem," *Scientific American*, February 2014.

Biedermannn, H., *Dictionary of Symbolism –Cultural Icons and the Meaning behind Them*, A Meridian Book, New York, 1994.

Bhadkamkar, A. and Fox, H., "Electron Charge Cluster Sparking in Aqueous Solutions," *Journal of New Energy*, Vol. 1, No. 4, 1996.

Blackwood, O.H., et al, *An Outline of Atomic Physics*, John Wiley & Sons, Inc., New York, 1955.

Bloyd, J.G., *Broken Arrow of Time - Rethinking the Revolution in Modern Physics*, Writers Club Press, San Jose, CA, 2001.

Born, M., *Atomic Physics*, Dover Publications, Inc., Mineola, NY, 1989.

Boscovich, R.J., *A Theory of Natural Philosophy*, The M.I.T. Press, Cambridge, MA, 1966.

Boslough, J., *Stephen Hawking's Universe - An Introduction to the Most Remarkable Scientist of Our Time*, Quill/William Morrow, New York, 1985.

Boslough, J., *Masters of Time - Cosmology at the End of Innocence*, Addison-Wesley Publishing Company, Reading, MA, 1992.

Bostick, W., "Mass, Charge, and Current: The Essence of Morphology," *Physics Essays*, Vol. 4, No. 1, pp. 45-59, March 1991.

Bowers, B., *Michael Faraday and Electricity*, Priory Press Ltd., London, 1974.

Brennan, R.P., *Heisenberg Probably Slept Here – The Lives, Times and Ideas of the Great Physicists of the 20th Century*, John Wiley & Sons, Inc., New York, 1997.

Broglie, L., de, *The Revolution in Physics*, The Noonday Press, New York, 1953.

Broglie, L., de, *New Perspectives in Physics*, Basic Books, Inc. Publishers, New York, 1962.

Burger, T.J., *Nature* **271**, 402, 1978.

C

Calladine, C.R. and Drew, H.R., *Understanding DNA – The Molecule and How It Works*, Second Edition, Academic Press, New York, 2002.

Cambier, J-L., at al, "Theoretical Analysis of the Electron Spiral Toroidal Concept," NASA/CR-2000-210654, Dec. 2000.

Capra, F., *The Web of Life*, Anchor Books/Random House, Inc., 1997.

Capra, F., *The Tao of Physics*, Shambhala Publications, Inc., 1999.

Carter, J., *The Other Theory of Physics - A Non-Field Unified Theory of Matter and Motion*, Absolute Motion Press, 2000.

Carrigan, Jr., R.A. and Trower, W.P.. *Particles and Forces at the Heart of the Matter,* W.H. Freeman and Company, New York, 1990.

Carroll, R.L., *The Energy of Physical Creation,* The Carroll Research Institute, P.O. Box 3425, Columbia, S.C., 29230, 1985.

Carter, J., *The Other Theory of Physics – A Non-Field Unified Theory of Matter and Motion,* Absolute Motion Press, Enumclaw, Washington, 2000.

Cartledge, P., *Democritus*, Routledge, New York, 1999.

Caspar, M., *Kepler*, Abelard-Schuman, London and New York, 1992.

Cecil, T.E. and Chern, S., *Tight and Taut Submanifolds*, Cambridge University Press, New York, 1997.

Chalidze, V., *Mass and Electric Charge in the Vortex Theory of Matter*, Universal Publishers, 2001.

Chen, C., at al, "Equilibrium and Stability Properties of Self-Organized Electron Spiral Toroid," Physics of Plasma, Volume 8, Number 10, October 2001.

Chen, Y., Editor, *Nanotubes and Nanosheets: Functionalization and Applications of Boron Nitride and Other Nanomaterials,* CRC Press, London, 2015.

Clark, G., *The Man Who Tapped the Secrets of the Universe*, The University of Science and Philosophy, Swannanoa, Waynesboro, 2000.

Clawson, C.C., *Mathematical Sorcery – Revealing the Secrets of Numbers*, Perseus Books, Cambridge, MA, 2001.

Clawson, C.C., *Mathematical Mysteries – The Beauty and Magic of Numbers*, Perseus Publishing, Cambridge, MA, 1999.

Close, F., *Neutrino,* Oxford University Press, 2010.

Close, F., Marten, M., & Sutton, C., *The Particle Explosion*, Oxford University Press, New York, 1994.

Coats, C., *Living Energies*, Gateway Books, Bath, UK, 1996.

CODATA Recommended Values, *The NIST Reference on Constants, Units and Uncertainties*, 2011.

Consa, O., "Helical Model of the Electron," The General Science Journal, June 2014.

Cook, N., *The Hunt for Zero Point - Inside the Classified World of Antigravity Technology*, Broadway Books, New York, 2001.

Cook, T.A., *The Curves of Life*, Dover Publications, Inc., New York, 1979.

Collins, H. and Pinch T., *The Golem – What Everyone Should Know about Science,* Cambridge University Press, 1993.

Compton, A., "The Size and Shape of the Electron," *Phys. Rev. Second Series*, vol. 14, no.3, pp 247-259, (1919).

Coxeter, H.S.M., *Introduction to Geometry*, John Wiley & Sons, Inc., New York, 1961.

Coxeter, H.S.M., *The Beauty of Geometry*, Dover Publications, Inc., New York, 1968.

Crandall, B.C., *Nanotechnology – Molecular Speculations on Global Abundance,* The MIT Press, Cambridge, Massachusetts, London, England, 1996.

Crew, H., *The Wave Theory of Light - Memoirs by Huygens, Young and Fresnel*, American Book Company, New York, 1900.

Cushing, J.T., *Philosophical Concepts in Physics – The History Relations between Philosophy and Scientific Theories,* Cambridge University Press, 1998.

D

Dalton, J., et al, *Foundations of the Atomic Theory: Comprising Papers and Extracts*, Alembic Club, Edinburgh, UK, 1968.

Davies, P.C.W. and Brown, J., *Superstrings - A Theory of Everything?,* Cambridge University Press, Cambridge, UK, 1988.

Davies, P. and Gribbin, J., *The Matter Myth - Dramatic Discoveries That Challenge Our Understanding of Physical Reality*, Touchstone Book/Simon & Schuster, New York, 1992.

Davies, P., *About Time - Einstein's Unfinished Revolution*, Simon & Schuster, New York, 1995.

Day, W., *Bridge from Nowhere - The Photonic Origin of Matter*, Rhombics, Cambridge, MA, 1996.

Day, W., *A New Physics - Foundation for New Directions*, Cambridge, MA, 2000.

Derbyshire, J., *Unkn( )wn Quantity - A Real and Imaginary History of Algebra*, A Plume Book, Published by Penguin Group, New York, 2007.

Di Mario, D., "Electrogravity: A Basic Link Between Electricity and Gravity," *Speculations in Science and Technology*, Vol. 20, No. 4, Dec. 1997.

Dibner, B., *Oersted - And the Discovery of Electromagnetism*, Blaisdell Publishing Company, New York, 1962.

Dijksterhuis, E.J., *Archimedes*, Princeton University Press, Princeton, N.J., 1987.

Dirac, PAM, "Quantized Singularities in the Electromagnetic Fields," Scribd.com. 1931-05-29.

Dixon, R., *Mathographics*, Dover Publications, Inc, New York, 1991.

Dmitriyev, V.P., "Mechanical Analogy for the Wave - Particle: Helix on Vortex Filament," *Apeiron*, Vol. 8, No. 2, April 2001.

Domb, C., *Clerk Maxwell and Modern Science - Six Commemorative Lectures*, The Athlone Press, University of London, UK, 1963.

Drake, S., *Galileo at Work, His Scientific Biography*, The University of Chicago Press, Chicago, 1978.

Dresselhaus, M.S. and Eklund, P.C., *Science of Fullerenes and Carbon Nanotubes: Their Properties and Applications*, Academic Press, 1996.

Drew, H.R., "The Electron as a Four-Dimensional Helix of Spin-1/2 Symmetry," *Physics Essays*, Vol. 12, No. 4, 1999.

Driscoll, R.B., *United Theory of Ether, Field and Matter*, Published by Author, Portland, OR, 1964.

Driscoll, R.B., *United Theory of Ether, Field and Matter (supplement)*, Published by Author, Oakland, CA, 1965.

Duncan, J.C., *Astronomy – A Textbook*, Fifth Edition, Harper & Brothers Publishers, New York, 1926.

E

Eckhart, L., *Four-Dimensional Space*, Indiana University Press, Bloomington, 1968.

Edwards, E.B., *Pattern and Design with Dynamic Symmetry*, Dover Publications, Inc., New York, 1932.

Edwards, L., *The Vortex of Life – Nature's Patterns in Space and Time*, Floris Books, Edinburgh, UK, 2006.

Ehrlich, R., *Crazy Ideas in Science - If You Might Even be True,* Princeton University Press, Prinston and Oxford, 2002.

Einstein, A. and Hopf, L., Ann. Phys., 33, 1096 (1910a): Ann. Phys., 33, 1105, 1910b.

Einstein, A., *Out of My Later Years*, A Citadel Press Book-Carol Publishing Group, New York, NY, 1991.

Einstein, A., *Relativity - The Special and the General Theory*, Crown Publishers, Inc., New York, 1961.

Einstein, A., Infeld, L., *The Evolution of Physics - From Early Concepts to Relativity and Quanta*, A Touchstone Book/Simon & Schuster, New York, 1966.

Einstein, A., "Aether and the Theory of Relativity," (Address on May 5, 1920, at the University of Leyden), *Journal of New Energy*, Vol. 7, No. 1, 2003.

Elgin, D., *The Living Universe – Where are We? Who are We? Where are We Going?*, Berrett-Koehler Publishers, Inc., San Francisco, CA, 2009.

Epstein, L.C., *Thinking Physics Is Gedanken Physics*, Insight Press, San Francisco, CA, 1983.

Epstein, L.C., *Relativity Visualized*, Insight Press, San Francisco, CA, 1992.

F

Farndon, J,. *The Great Scientists – From Euclid to Stephen Hawking,* Metro Books, New York, 2007.

Farrington, B., *Greek Science - Its Meaning for Us*, Penguin Books, Baltimore, MD, 1971.

Feber, A., "Supertwistors and Conformal Supersymmetry," *Nuclear Physics B* **132**: 55-64, 1978.

Ferguson, K., *Stephen Hawking - Quest for a Theory of Everything*, Bantam Books, New York, 1992.

Feynman, R.P., *Six Easy Pieces and Six Not-So-Easy Pieces,* Perseus Publishing, Cambridge, Massachusetts, 1995.

Flander, T.V., *Dark Matter, Missing Planets & New Comets – Paradoxes Resolved, Origins Illuminated,* Revised Edition, North Atlantic Books, Berkley, California, 1993.

Flood, R. and Lockwood, M., *The Nature of Time*, Basil Blackwell, Inc., Cambridge, MA, 1990.

Folger, T., "Tangled Up In Strings – Two Books Say That Today's Theoretical Physicists Are Way Off Course," *Discover*, Sept. 2006.

Ford, K.W., *101 Quantum Questions*, Harvard University Press, Cambridge, Massachusetts, 2011.

Fowler, P.W. and Manolopoulos, D.E., *An Atlas of Fullerenes*, Dover Publications, 2007.

Frank, P., *Einstein - His Life and Times*, Da Capo Press, Inc., New York, 1947.

Fraser, et al, *The Search for Infinity*, Facts on File, Inc., New York, 1995.

Freedman, D.H., "The Mysterious Middle of the Milky Way," *Discover*, November 1998.

Freeman, K. and McNamara, G., *In Search of Dark Matter*, Springer Praxis Publishing, Chichester, UK, 2006.

Friedman, N., *Bridging Science and Spirit - Common Elements in David Bohm's Rhysics, The Perennial Philosophy and Seth*, Living Lake Books, St. Louis, MO, 1994.

Fritzsch, H., *Quarks - The Stuff of Matter*, Basic Books, Inc., New York, 1983.

Fritzsch, H., *The Creation of Matter - The Universe From Beginning to End*, Basic Books, Inc., New York, 1984.

Funk & Wagnalls New Encyclopedia, Funk & Wagnalls, Inc., USA, 1966.

G

Gamow, G., *The Great Physicists from Galileo to Einstein,* Dover Publications, Inc., New York, 1988.

Gamow, G., *Thirty Years That Shook Physics - The Story of Quantum Theory*, Dover Publications, Inc., New York, 1985.

Gamow, G., *One, Two, Three ... Infinity - Facts and Speculations of Science*, Dover Publications, Inc., New York, 1988.

Gardner, M., *New Mathematical Diversions from Scientific American*, Simon and Schuster, New York, 1966.

Gardner, M., *Knotted Doughnuts and Other Mathematical Entertainments*, W.H. Freeman and Company, New York, 1986.

Gasperini, M., *The Universe Before the Big Bang,* Springer, Berlin, 2010.

Gauthier, R., "Faster-than-light quantum models of the photon and the electron", in M. S. El-Genk, *(*ed.) *"Space Technology and Applications International Forum – STAIF 2007"*, American Institute of Physics 978-0-7354-0386-4/07, p1099-1108, 2007.

Gauthier, R., "Transluminal energy quantum models of the photon and the electron", in R.L. Amoroso, P. Rowlands & L.H. Kauffman (eds.) *The Physics of Reality: Space, Time, Matter, Cosmos, 8th Symposium in Honor of Mathematical Physicist Jean-Pierre Vigier*, Hackensack: World Scientific, 2013.

Gauthier, R., "A transluminal energy quantum model of the cosmic quantum", in R.L. Amoroso, P. Rowlands & L.H. Kauffman (eds.) *The Physics of Reality: Space, Time, Matter, Cosmos, 8th Symposium in Honor of Mathematical Physicist Jean-Pierre Vigier*, Hackensack: World Scientific, 2013.

Gautreau, R. and Savin, W., *Schaum's Outline of Theory and Problems of Modern Physics,* McGeaw-Hill, New York, 1978.

Gazale, M.J., *Gnomon*, Princeton University Press, Princeton, N.J., 1999.

Geerlings, G.K., *Wrought Iron In Architecture – An Illustrated Survey,* Dover Publications, Inc, New York, 1983.

Gell-Mann, M., *Quark and the Jaguar – Adventures in the Simple and the Complex*, A W, H. Freeman/Owl Book, Henry Holt and Company, LLC, New York, 1994.

Gell-Mann, M., *Complexity*, Vol. 1, no 5, John Wiley and Sons, Inc., New York, 1995/96.

Genz, H., *Nothingness - The Science of Empty Space*, Perseus Books, Reading, MA, 1999.

Geymonat, L., *Galileo Galilei: A Biography and Inquiry Into His Philosophy of Science*, McGraw-Hill Book Company, 1965.

Ghosh, A., *Origin of Inertia*, Apeiron, Montreal, Canada, 2000.

Ghyka, M., *The Geometry of Art and Life*, Dover Publications, Inc, New York, 1977.

Gillispie, C.C., *Dictionary of Scientific Biography*, Vol. IV, Charles Scribner's Sons, New York, 1971.

Ginzburg, V.L., *Theoretical Physics and Astrophysics*, Pergamon Press, 1979.

Ginzburg, V.L., *Physics and Astrophysics. A Selection of Key Problems*, Pergamon Press, 1985.

Gleick, J., *Chaos - Making a New Science*, Penguin Books, New York, 1988.

Gleick, J., *Genius - The Life and Science of Richard Feynman*, Pantheon Books, New York, 1992.

Gorini, C.A., *Geometry*, Facts On File, Inc., New York, 2003.

Goswami, A., *The Self-Aware Universe - How Consciousness Creates the Material World*, Penquin Putnam Inc., 1993.

Graver, J.E., "The Structure of Fullerene Signatures," *DIMACS Series in Discrete Mathematics and Theoretical Computer Science*, American Mathematical Society, 2005.

Gray, A., *Lord Kelvin - An Account of His Scientific Life and Work*, E.P. Dutton & Co., 1908.

Gray, A., *Modern Differential Geometry of Curves and Surfaces*, CRC Press, Boca Raton, 1993.

Greene, B., *The Elegant Universe - Superstrings, Hidden Dimensions, and the Quest for the Ultimate Theory*, W.W. Norton & Company, New York, 1999.

Greene, B., *The Fabric of the Cosmos*, Alfred A. Knopf, New York, 2004.

Gribbin, J., *Q Is For Quantum, An Encyclopedia of Particle Physics*, A Touchstone Book/ Simon & Schuster, New York, 2000.

Gribbin, J., *Schrödinger's Kittens and the Search for Reality,* Little, Brown & Company (Canada) Limited, 1995.

Guillen, M., *Five Equations that Changed the World*, Hyperion, New York, 1995.

H

Haisch, B., *The Purpose-Guided Universe – Believing in Einstein, Darwin, and God*, The Career Press, Inc., Franklin Lakakes, NJ, 2010.

Haisch, B., *The God Theory - Universes, Zero-Point Fields, and What's Behind It All*, Weser Books, San Francisco, CA, 2006..

Hall, A.R., *Isaac Newton - Adventurer in Thought*, Blackwell Publishers, Oxford, UK, 1994.

Hambidge, J., *Practical Applications of Dynamic Symmetry*, The Devin-Adair Company, New York, 1967.

Hargittai, I., Pickover, C.A., Editors, *Spiral Symmetry*, World Scientific Publishing Co., Singapore, 1992.

Harrington, P.S, *Star Watch*, John, Wiley & Sons, Inc., Hoboken New Jersey, 2003.

Harris, P.J.F., *Carbon Nanotube Science: Synthesis, Properties and Applications,* Cambridge University Press, Cambridge, UK, 2011.

Harrison, L.P., *Meteorology*, National Aeronautics Council, Inc., New York, 1942.

Hatch, E., *Modern Physics from a Classical Scale Perspective- Part 1 Concept Confirmed,* RWWAA Publication, Auburn, California, 2004.

Hawking, S., *Black Holes and Baby Universes and Other Essays*, Bantam Books, New York, 1993.

Hawking, S., *A Brief History of Time - From the Big Bang to Black Holes*, Bantam Books, New York, 1990.

Hawking, S., *On the Shoulders of Giants,* Running Press, Philadelphia, London, 2004.

Hawking, S. and Penrose, R., *The Nature of Space and Time*, Princeton University Press, Princeton, NJ, 2000.

Heath, J.L., *The Works of Archimedes*, Dover Publications, Inc., New York, 1953.

Heilbron, J.L., *The Dilemmas of an Upright Man - Max Planck as Spokesman for German Science*, University of California Press, Berkeley, 1986.

Heisenberg, E., *Inner Exile - Recollections of a Life with Werner Heisenberg*, Birkhauser Boston, MA, 1980.

Helmholtz, H., *On the Sensation of Tone – As a Physical Basis for the Theory of Music,* Dover Publications, New York, 1954.

Henderson, L.D., *The Fourth Dimension and Non-Euclidean Geometry in Modern Art*, Princeton, 1983.

Herzberg, G., *Atomic Spectra and Atomic Structure*, Dover Publications, Inc., New York, 1944.

Hey, N., *Solar System*, Weidenfeld & Nicolson, The Orion Publishing Group, Wellington House, London, UK.

Hey, T. and Walters, P., *The New Quantum Universe*, Cambridge University Press, UK, 2003.

Hippel von, F., *Citizen Scientist*, A Touchstone Book/Simon & Schuster, New York, 1991.

Hoffmann, B., *Albert Einstein - Creator and Rebel*, New American Library, New York, 1972.

Hoffman, R.N., "Controlling Hurricanes – Can Hurricanes and Other Severe Tropical Storms be Moderated or Deflected?" *Scientific American*, Oct. 2004.

Hooft, G., *In Search of the Ultimate Building Blocks*, Cambridge University Press, UK, 1997.

Horgan, J., "Gravity Quantized? - A Radical Theory of Gravity Weaves Space From Tiny Loops," *Scientific American*, September 1992.

Hotson, D.L., "Dirac's Equation and the Sea of Negative Energy, Part 1," *Infinite Energy*, Vol. 8, Issue 43, 2002.

Hotson, D.L., "Dirac's Equation and the Sea of Negative Energy, Part 2," *Infinite Energy*, Vol. 8, Issue 44, 2002.

Hurley, W.M., *Prehistoric Cordage – Aldine Manuals on Archeology 3,* Taraxacum, Washington, 1979.

I

Icke, V., "From Expansion to Intelligence in the Universe," *Speculations in Science and Technology*, Vol. 14, No. 4, 1991.

Ipsen, D.C., *Archimedes: Greatest Scientist of the Ancient World*, Enslow Publishers, Inc., Hillside, N.J., 1988.

J

Jammer, M., *Concepts of Mass in Classical and Modern Physics*, Dover Publications, Inc., Mineola, New York, 1997.

Jammer, M., *Concepts of Force*, Dover Publications, Inc., Mineola, New York, 1999.

Jean, Sir, J, *Science & Music,* Dover Publications, Inc., New York, 1968

Johnson, G., "The Inelegant Universe – Two New Books Argue That It Is Time For String Theory To Give Way," *Scientific American*, September, 2006.

Jefimenko, O.D., *Gravitation and Cogravitation*, Electret Scientific Company, Star City, West Virginia, 2006.

Jones, B.Z., *The Golden Age of Science,* Simon and Schuster, New York, 1966.

Jonsson, I., *Emanuel Swedenborg*, Twayne Publishers Inc., New York, 1971.

K

Kafatos, M. and Nadeau, R., *The Conscious Universe - Part and Whole in Modern Physical Theory*, Springer-Verlag New York, Inc., New York, 1990.

Kaku, M., *Physics of the Impossible,* Anchor Book, A Division of Random House, Inc., 2008.

Kaku, M., *Visions – How Science Will Revolutionize the 21st Century,* Anchor Books, Doubleday, New York, London, 1997.

Kaku, M., *Beyond Einstein - The Cosmic Quest for Theory of the Universe*, Anchor Books/Doubleday, New York, 1995.

Kaku, M., *Hyperspace*, Anchor Books/Doubleday, New York, 1994.

Kaku, M., *Introduction to Superstrings*, Springer-Verlag, New York, 1988.

Kanarev, F.M., "Model of the Electron," APERON, Vol. 7. Nr. 3-4, July-October, 2000.

Kanarev, F.M., *The Foundation of Physchemistry of Microworld*, Kuban State Agrarian University (KSAU), Krasnodar, Russia, 2002.

Kane, G., *The Particle Garden - Our Universe as Understood by Particle Physicists*, Addison-Wesley Publishing Company, Reading, MA, 1995.

Kanigel, R., *The Man Who Knew Infinity - A life of the Genius Ramanujan,* Washington Square Press, Published by Pocket Books, New York, London, 1991.

Kaplan, R., *The Nothing That Is - A Natural History of Zero*, Oxford University Press, Oxford, New York, 1999.

Kaplan, R. and Kaplan, E., *The Art of The Infinite*, Oxford University Press, Oxford, New York, 2003.

Kaufmann, W.J., *Black Holes and Warped Spacetime*, W.H. Freeman and Company, San Francisco, CA.

Kimura, Y.G., *The Book of Balance*, (Translation), The University of Science and Philosophy, Contact Printing, North Vancouver, B.C., Canada, 2002.

Kimura, Y.G., "The Transcendent Unity of Science and Spirituality," *VIA – Vision in Action*, Vol. 2, No. 1 & 2, 2004.

King, M.B., "Vortex Filaments, Torsional Fields and the Zero-Point Energy," *Journal of New Energy*, Vol. 3, No. 2/3, 1998.

King, M.B., "Dual Vortex Forms: The Key to a Large Zero-Point Energy Coherence," *Journal of New Energy*, Vol. 5, No. 2, 2000.

King, M.B., *Quest for Zero Point Energy*, Adventures Unlimited Press, Kempton, IL, 2001.

King, M.B., *Tapping the Zero-Point Energy,* Paraclete Publishing, Provo, Utah, 1989.

Knight, D.C., *The Science Book of Meteorology*, Franklin Watts, Inc., New York, 1964.

Krauss, L.M., *Quintessence - The Mystery of Missing Mass in the Universe*, Basic Books, New York, NY, 2000.

Kumar, S., "A Spiral Structure for Elementary Particles," Int. J. Res. Vol. 1, Issue 6, July 2014.

Kumar, S., "Journey of the Universe from Birth to Rebirth with Insight into Unified Interaction of Elementary Particles with Spiral Structure," Int. J. Res. Vol. 1, Issue 9, October 2014.

Kumar, S., "Quantum Spiral Theory," Int. J. Res. Vol. 2, Issue 1, January 2015.

Kumar, S., "Spiral Hashed Information Vessel," International Journal of Scientific & Engineering Research, Vol. 4, Issue 6, April 2015.

Kumar, S., "Spiral Structure of Elementary Particles Analogous to Sea Shells: A Mathematical Description," International Journal of Current Research, Vol. 7, Issue 02, Feb. 2015, p. 12814.

Kumar, S., "Mass-Energy Equivalence in Spiral Structure for Elementary Particles and Balance of Potentials," International Journal of Scientific & Engineering Research. Vol. 6, Issue 6, July 2015.

L

Lamb, G.L., "Solutions and the Motion of Helical Curves," *Physical Review Letters*, Vol. 37, No. 5, August 1976.

Lakhtakia, A. and Weiglhofer, W.S., "Time-Dependent Beltrami Fields in Free Space: Dyadic Green Functions and Radiation Potentials," *Physical Review E*, Vol. 49, Number 6, June 1994.

Lakhtakia, A. and Weiglhofer, W.S., "Covariances and Invariances of the Beltrami-Maxwell Postulates," *IEE Proc. - Sci. Meas. Technol.*, Vol. 142, No. 3, May 1995.

Lang, T.G., "Proposed Unified Field Theory – Part II: Protons, Neutrons and Fields," *Galilean Electrodynamics*, Vol. 12, No. 6, Nov./Dec. 2001.

Larsen, R., et al, *Emanuel Swedenborg - A Continuing Vision*, Swedenborg Foundation, Inc., New York, 1988.

Laugwitz, D., *Differential and Riemannian Geometry*, Academic Press, New York, 1965.

Lauwerier, H., *Fractals - Endlessly Repeated Geometrical Figures*, Princeton University Press, Princeton, New Jersey, 1991.

Lederman, L.M. and Teresi, D., *The God Particle*, Bantam Doubleday Dell Publishing Group, Inc., New York, 1993.

Lederman, L.M. and Hill, C.T., *Symmetry and the Beautiful Universe*, Prometheus Books, New York, 2004.

Lederman, L.M. and Hill, C.T., *Quantum Physics for Poets*, Prometheus Books, New York, 2011.

Lederman, L. and Hill, C., *Beyond the God Particle*, Prometheus Books, New York, 2013.

Lerner, E.J., *The Big Bang Never Happened*, Vintage Books, Random House, Inc., New York, 1992.

Lewis, H., *Geometry – A Contemporary Course,* Third Edition, McCormick-Mathers Publishing Company, Cincinnati, Ohio, 1973.

Lewis, J.R., *Scientology*, Cary, NC, Oxford University Press, 2009.

Lindgren, C.E., *Four-Dimensional Descriptive Geometry*, McGraw-Hill Book Company, New York, 1968.

Lindley, D., *The End of Physics - The Myth of a United Theory*, HarperCollins Publishers, Inc., 1993.

Lipschultz, M.M, *Differential Geometry,* Schaum's Outline Series, McGraw-Hill, New York, 1969.

Livio, M., *The Equation That Couldn't Be Solved – How Mathematical Genius Discovered the Language of Symmetry*, Simon and Schuster, New York, 2005.

Livio, M., *The Golden Ratio*, Broadway Books, New York, 2002.

Lockwood, E.H., *A Book of Curves*, Cambridge University Press, New York, 1961.

Lomberg, J., *Unified Force Theory, Dark Matter and Consciousness,* The Aenor Trust, PO Box 4706, Salem, Oregon, 2004.

Lorentz, H.A., *Problems of Modern Physics - A Course of Lectures Delivered in the CA Institute of Technology*, Ginn and Company, Boston, 1927.

Lucas, C.W., "A Classical Electromagnetic Theory of Elementary Particles," *Journal of New Energy*, Vol. 6, No. 4, 2002.

Lucas, C.W. "A Classical Electromagnetic Theory of Elementary Particles Part 2, Interwining Charge-Fibers," The Journal of Common Sense Science, Foundation of Science, May 2005, Vol. 8 No. 2.

Ludwig, C., *Michael Faraday - Father of Electronics*, Herald Press, Scottdale, PA, 1978.

Lugt, H.J., *Vortex Flow in Nature and Technology*, Krieger Publishing Company, Malabar, Florida, 1995.

Lykken, J. and Spiropulu, M., "Supersymmetry and the Crisis in Physics," *Scientific American*, May 2014.

M

Maldacena, J., "The Illusion of Gravity," *Scientific American*, pp. 57-63, November 2005.

Magueijo, J., *Faster Than the Speed of Light - The Story of A Scientific Speculation,* Penguin Books, New York, 2003.

Manning, J., *The Coming Energy Revolution - The Search for Free Energy*, Avery Publishing Group, Garden City Park, New York, 1996.

Manning, H.P., *The Fourth Dimension Simply Explained*, Dover Publications, Inc., New York, 1960.

Maor, E., *e: The Story of a Number*, Princeton University Press, Princeton, NJ, 1994.

Maor, E., *Trigonometric Delights*, Princeton University Press, Princeton, NJ, 1998.

Marsden, J.E. and McCracken, M., *The Hopf Bifurcation and Its Applications*, Springler-Verlag, New York, Berlin, 1976.

Mazur, B., *Imagining Numbers (particularly the square root of minus fifteen)*, Picador, New York, 2003.

McCrea, W.H., "Arthur Stanley Eddington," *Scientific American*, June 1991.

McCutcheon, M., *The Final Theory – Rethinking Our Scientific Legacy*, Universal Publishers, Boca Raton, Florida, 2004.

McLeish, J., *The Story of Numbers*, Fawcett Columbine, New York, 1991.

Meacher, M., *Destination of the Species – The Riddle of Human Existence*, Books, Winchester, UK, Wasington USA, 2009.

Melker, A.A. and Krupina, M.A., "Designing Muni-Fullerences and Their Relatives on Graph Basis," *Materials Physics and Mechanics 20*, 18-24 2014.

Messent, J., *Embroidery & Architecture,* B.T Batsford Ltd., London, 1985.

Millar, D., et al, The Cambridge Dictionary of Scientists, Cambridge University Press, 1996.

Miller, A.I., *137 – Jung, Pauli and the Pursuit of the Scientific Obsession*, W.W Norton & Company, Inc., New York, 2009.

Miller, A.I., *Albert Einstein's Special Theory of Relativity,* Springer-Verlag, New York, Berlin, 1997.

Milton, R., *Alternative Science - Challenging the Myths of the Scientific Establishment*, Park Street Press, Rochester, Vermont, 1996.

Mitchell, W.C., *Bye Bye Bing Bang – Hello Reality,* Cosmic Sense Books, Carson City, Nevada, 2002.

Mitsopoulos, T.D., "Similarity Between Elementary Particles and Electric Circuits," *Galilean Electrodynamics*, Vol. 12, No. 6, Nov./Dec. 2001.

Moore, W., *Schrodinger - Life and Thought*, Cambridge University Press, UK, 1992.

Mortimer, S., *Techniques of Spiral Work – A Practical Guide to the Craft of Making Twists by Hand,* Linden Publishing, Fresno, California, 1995.

Moyer, M., "Is the Space Digital," *Scientific American*, February 2012.

Mugnai, D., et al, "Observation of Superluminal Behaviors in Wave Propagation," *Physical Review Letters*, Vol. 84, Number 21, May 2000.

Murchie, G., *The Seven Mysteries of Life - An Exploration in Science and Philosophy*, Houghton Mifflin Company, Boston, 1978.

N

Nahin, P.J., *An Imaginary Tale – The Story of* $\sqrt{-1}$, Princeton University Press, Princeton, New Jersey, 1998.

Nakahara, M., *Geometry, Topology and Physics,* Second Edition, Taylor &Francis, Taylor & Francis Group, New York, London, 2003.

Nash, C. and Sen, S., *Topology and Geometry for Physicists*, IBI Global, London, UK, 1983.

Nernst, W., Verh. Dtsch. Phys. Ges., 18, 83, 1916.

Newton, I., *The Principia*, Prometheus Books, Amherst, New York, 1995.

Nierengarten, J-F., Editor, *Fullerenes and Other Carbon-Rich Nanostructures (Structure and Bonding)*, Springer, 2014.

Niven, W.D., *The Scientific Papers of James Clerk Maxwell*, Dover Publications, Inc., New York, 1890.

Novikov, I.D., *The River of Time*, Cambridge University Press, Cambridge, UK, 1998.

O

Okun, L.B., "The Concept of Mass," *Physics Today*, Vol. 42, June 1989.

Oliwensrein, L., "Bent out of Shape," *Discover*, July 1993.

Oros di Bartini, R., "Relations Between Physical Constants," *Progress in Physics*, v. 3, pp. 34-40, October 2005.

Oros di Bartini, R., "Some Relations Between Physical Constants," *Doklady* Acad. Nauk USSR, v. 163, No. 4, pp. 861-864, 1965.

Oschman, J.L. and Schman N.H., "Vortical Structure of Light and Space: Biological Implications," *J Vortex Sci Technol.* 2:1, 2015.

Oschman, J.L. and Schman N.H., "The Heart as a Bi-Directional Scalar Field Antena," *J Vortex Sci Technol.* 2:2, 2015.

Oschman, J.L., *Energy Medicine – The Scientific Basis,* Second Edition, Elsevier, New York, 2016.

P

Pagels, H.R., *The Cosmic Code – Quantum Physics as the Language of Nature*, Bantam Books, New York, 1982.

Panek, R., *The 4% Universe – Dark Matter, Dark Energy, and the Race to Discover the Rest of Reality*, Houghton Mifflin Harcourt, Boston, New York, 2011.

Pappas, T., *The Joy of Mathematics – Discovering Mathematics All Around You,* Wide World Publishing, Tetra, 1989.

Parry, A., *The Russian Scientist*, The Macmillan Company, New York, 1973.

Parson, A.L., "A Magneton Theory of the Structure of the Atom," Smithsonian Miscellaneous Collections, Vol. 65, No. 11, Publication 2371, Nov. 29, 1915.

Pauli, W., *Theory of Relativity*, Pergamon Press, London, UK, 1958.

Peat, F.D., *Superstrings and the Search for The Theory of Everything*, Contemporary Books, Lincolnwood (Chicago), ILL,1988.

Pedoe, D., *Geometry - A Comprehensive Course*, Dover Publications, Inc., New York, 1988.

Peebles, P.J.E., *Principles of Physical Cosmology*, Princeton University Press, Princeton, New Jersey, 1993.

Penrose, R., "Twistor Quantization and Curved Space-time." *International Journal of Theoretical Physics* (Springer Netherlands), **1**: 61-99, 1968.

Penrose, R., *The Road to Reality - A Complete Guide to the Laws of the Universe*, Alfred A. Knopf, New York, 2005.

Penrose, R., *Shadows of the Mind – A Search for the Missing Science of Consciousness,* Oxford University Press, 1994.

Peratt, A.L., "Birkeland and the Electromagnetic Cosmology," *Sky & Telescope*, May 1985.

Physical Review D: Particles and Fields, Vol. 54, The American Physical Society, 1996.

Pierson, H.O., *Handbook of Carbon, Graphite, Diamond and Fullerenes: Properties and Applications (Material Science and Process Technology)*, Noyes Publications, 1994.

Pickover, C.A., *Mathematics and Beauty II; Spirals and "Strange" Spirals in Civilization, Nature, Science, and Art*, IBM Thomas J. Watson Research Center, Yorktown Heights, NY, 1987.

Polyakov, A., "Gauge Fields and Strings," Harwood Academic Publishers 1987, *Nucl. Phys.* **B396**, 367, 1993.

Ponomarev, C.D. and Andreeva, L.E., *The Calculation of Elastic Elements of Machines and Sensors,* Machinostroenie, Moscow, 1980.

Porter, R., *The Biographical Dictionary of Scientists*, Oxford University Press, New York, 1994.

Posamentier, A.S. and Lehmann, I., *A Biography of the World's Most Mysterious Number*, Prometheus Books, Amherst, New York, 2004.

Potemra, T. A., "Hannes Alfven, Father of Space Plasma Physics," Geomagne-tism and Aeronomy with Special Historical Case Studies, IAGA Newsletters 29/1997, Published by IAGA, Germany, p.101, 1997.

Price, W.C., et al, *Wave Mechanics; The First Fifty Years - A Tribute to Professor Louis De Broglie*, John Wiley & Sons, New York-Toronto, 1973.

Price, H., *Time's Arrow and Archimedes' Point*, Oxford University Press, New York, 1996.

Purce, J., *The Mystical Spiral- Journey of the Soul,* Thames and Hudson, 1974.

Purdy, S., and Sandak, C.R., *Ancient Greece*, Franklin Watts, New York, 1982.

Puthoff, H.E., et al, "Engineering the Zero-Point Field and Polarizable Vacuum for Interstellar Flight," *Journal of New Energy*, Vol. 6, No. 1, 2001.

R

Randless, J. *Breaking the Time Barrier*, Paraview Pocket Books, New York, London, 2005,

Reed, D., "Excitation and Extraction of Vacuum Energy Via EM-Torsional Field Coupling - Theoretical Model," *Journal of New Energy*, Vol. 3, No. 2/3, 1998.

Reed, D., "A New Paradigm for Time – Evidence From Empirical and Esoteric Sources," *Journal of New Energy*, Vol. 6, No. 2, 2001.

Resnick, R., *Introduction to Special Relativity,* Jon Wiley & Sons, Inc., New York, London, 1968.

Ridley, B.K., *Time, Space and Things*, Cambridge University Press, Cambridge, UK, 1994.

Riordan, M., *The Hunting of the Quark - A True Story of Modern Physics*, Simon and Schuster/Touchstone, New York, 1987.

Riordan, M. and Schramm, D.N., *The Shadows of Creation - Dark Matter and the Structure of the Universe*, W.H. Freeman and Company, New York, 1991.

Rucker, R., *The Fourth Dimension*, Houghton Mifflin Company, Boston, 1984.

Russell, P., *The White Hole in Time*, Harper San Francisco, 1992.

Russell, W. *The Universal One*, University of Science and Philosophy, Swannanoa, Waynesboro, Virginia, 1974.

Russell, W., *A New Concept of the Universe*, The University of Science and Philosophy, Swannanoa, Waynesboro, VA, 1989.

Russell, W., *The Secret of Light*, The University of Science and Philosophy, Swannanoa, Waynesboro, VA, 1994.

Ryu, C., *The Grand Unified Theory – A Scientific Theory of Everything*, PublishAmerica, Baltimore, 2004.

S

Saito, R., Author, Editor, *Physical Properties of Carbon Nanotubes*, Imperial College Press, London, 1998.

Salem, K.G., *The New Gravity - A New Force - A New Mass - A New Acceleration - Unifying Gravity with Light*, Salem Books, Johnstown, PA, 1994.

Sagan, C., *Cosmos,* Ballantine Books, New York, 1980.

Sanders, P.A. Jr., *Scientific Vortex Information*, Free Soul Publishing, Sedona, AZ, 1992.

Sano, C., "Twisting & Untwisting of Spirals of Aether and Fractal Vortices Connecting Dynamic Aethers," *Journal of New Energy*, Vol. 6, No. 2, 2001.

Sarg, S., "A Physical Model of the Electron According to the Basic Structure of Matter Hypothesis," *Physics Essays*, Vol. 16, No.2, 180-195, 2003.

Sarg, S, "Basic Structure of Matter – Supergravitation Unified Theory Based on an Alternative Concept of the Physical Vacuum," Proceedings of the 17th Annual Conference of the NPA at Long Beach, CA, Vol. 7, pp. 479-484, 23-26 June, 2010.

Savov, E., *Theory of Interaction - The Simplest Explanation of Everything*, Geones Books, Sofia, Bulgaria, 2002.

Schneider, M.S., *A Beginner's Guide to Constructing the Universe*, HarperPerennial, New York, 1995.

Schweighauser, C.A., *Astronomy from A to Z – A Dictionary of Celestial Objects and Ideas,* Sangamon State University, Springfield, Illinois, 1991.

Schwenk, T., *Sensitive Chaos – The Creation of Flowing Forms in Water and Air,* Rudolf Steiner Press, Hillside House, East Sussex, 2008.

Schwerdtfeger, H., *Geometry of Complex Numbers – Circle Geometry, Moebius Transformation, Non-Euclidean Geometry*, Dover Publication, Inc., New York, 1979.

Scientific American, *The Enigma of Weather*, Scientific American, New York, NY, Oct. 2004.

Segal, V.M., "Materials Processing by Simple Shear," *Mat. Sci & Eng.,* vol. 197, 157-164, 1995.

Segal, V.M., "Equal Channel Angular Extrusion: From Macro Mechanics to Structure Formation," *Mat. Sci & Eng.,* vol. 271, 322-333, 1999.

Segal, V.M., "Severe Plastic Deformation: Simple Shear versus Pure Shear," vol. 338, pp. 331-344, 2002.

Seggern, D.H. von, *CRC Handbook of Mathematical Curves and Surfaces*, CRC Press, Boca Raton, Florida, 1990.

Seggern, D.H. von, *CRC Standard Curves and Surfaces*, CRC Press, Boca Raton, Florida, 1993.

Segre, E., *Nuclei and Particles - An Introduction to Nuclear and Subnuclear Physics*, W.A. Benjamin, Inc., New York, 1965.

Seife, C., *Zero - The Biography of a Dangerous Idea*, Viking Penquin, New York, 2000.

Semat, H., *Introduction to Atomic and Nuclear Physics,* Rinehart & Company, Inc., New York, 1958.

Series, G.W., *Advances - The Spectrum of Atomic Hydrogen*, World Scientific, New Jersey, 1988.

Serway, R.A., *Physics for Scientist & Engineers with Modern Physics*, 3rd Edition, Saunders Golden Sunburst Series, Saunders College Publishing, Philadelphia, PA, 1990.

Seward, C., "Ball Lightning Events Explained as Self-Stable Spinning High-Density Plasma Toroids or Atmospheric Spheromaks," IEEE *Access* Practical Innovations, Volume 2, 2014, 153-59.

Sharlin, H.I., *Lord Kelvin - The Dynamic Victorian*, The Pennsylvania State University Press, PA, 1979.

Sheka, E., *Fullerences: Nanochemistry. Nanomagnetism, Nanomedicine, Nanophotonics,* CRC Press, 2011.

Siegfried, T., *Strange Matters – Undiscovered Ideas at the Frontiers of Space and Time,* The Berkley Publishing Group, A division of Penguin Group, New York, 2004.

Siegfried, T., *The Bit and the Pendulum*, John Wiley & Sons, Inc., New York, 2000.

Simhony, M., *Invitation to the Natural Physics of Matter, Space, Radiation*, World Scientific, New Jersey, London, 1994.

Smolin, L., *The Trouble With Physics: The Rise of String Theory, The Fall of a Science, and What Comes Next,* Houghton Mifflin, 2006.

Sprott, J.C., *Strange Attractions - Creating Patterns in Chaos*, M&T Books, New York, 1993.

Sproull, R.L., *Modern Physics – A Textbook for Engineers,* John Wiley & Sons, New York, 1956.

Stenger, V.J., *God and the Atom – From Democritus to the Higgs Boson: The Story of a Triumphant Idea*, Prometheus Books, New York, 2013.

Sternberg, S., *Curvature in Mathematics and Physics*, Dover Publications, Inc., New York, 2012.

Sternglass, E.J., *Before the Big Bang - The Origins of the Universe*, Four Walls Eight Windows, New York, NY, 1997.

Strogatz, S., *Sync - The Emerging Science of Spontaneous Order*, Hyperion, New York, 2003.

Sunden, O., "Time-Space-Oscillation: Hidden Mechanism Behind Physics," *Galilean Electrodynamics*, Vol. 12, Special Issue 2, Fall 2001.

Swedenborg, E., *The Principia*, Swedenborg Society, London, 1912.

Synge, J.L., "The Electrodynamic Double Helix," In *Magic Without Magic: John Archibald Wheeler* - A Collection of Essays in Honor of His Sixtieth Birthday, edited by John R. Klauder, W. H. and Company, San Francisco, 1972.

T

Tanaka, K., Editor, Iijima, S., *Carbon Nanotubes and Graphene, Second Edition,* Nanotube Research Center, Tsukuba, Japan, 2014.

Talbot, M., *The Holographic Universe*, Harper Perennial, 1992.

Tewari, P., *Universal Principles of Spacetime and Matter - A Call for Conceptual Reorientation,* Crest Publishing House, New Deli, 2002.

Tewari, P., "On the Space-Vortex Structure of the Electron," www.tewari.org/ Theory_Papers/Tewari-Final%20Proof.pdf. 2005.

Thomson, J.J., *A Treatise on the Motion of Vortex Rings*, MacMillan and Co., London, 1883.

Thomson, J.J., *Electricity and Matter*, Charles Scribner's Sons, New York, 1904.

Thomson, D.W. and Bourassa, J.D., *Secrets of Aether*, Published by The Aenor Trust, Salem, OR, 2004.

Time-Life Books, *A Soaring Spirit - Time Frame BC 600-400*, The Time Inc. Book Company, Alexandria, VA, 1987.

Time-Life Books, *Empires Ascendant - Time Frame 600 BC - AD 200*, The Time Inc. Book Company, Alexandria, VA, 1987.

Tricker, R.A., *The Contributions of Faraday and Maxwell to Electrical Science*, Pergamon Press Ltd., London, UK, New York, 1987.

Thorne, K.S., *Black Holes & Time Warps - Einstein's Outrageous Legacy*, W.W. Norton & Company, New York, 1994.

Treasures of Early Irish Art: 1500 B.C. to 1500 A.D., From the Collections of the National Museum of Ireland, Royal Irish Academy, Trinity College, Dublin, 1977.

U

Unger, R.M. and Smolin, L., *The Singular Universe and the Reality of Time*, Cambridge University Press, Cambridge, UK, 2015.

V

Von Stade, S., "Owner". *Flowtoys.* Flowtoys (Toroflux).

Valens, E.G., *The Attractive Universe: Gravity and Shape of Space*, Motion, Magnet, 1969.

Van der Laan, C., "The Vortex Theory of Atoms," Thesis for the Master's Degree in History and Philosophy of Science, Institute for History and Foundation of Science, Utrecht University, Dec. 2012.

Van Eenwik, J.R., *Archetypes & Strange Attractors - The Chaotic World of Symbols*, Inner City Books, Toronto, Canada, 1997.

Van Flandern, T., *Dark Matter, Missing Planets & New Comets - Paradoxes Resolved, Origins Illuminated*, North Atlantic Books, Berkeley, CA, 1993.

Valone, T., "Inside Zero Point Energy," *Journal of New Energy*, Vol. 5, No. 4, 2001.

Van Nostrand's Scientific Encyclopedia, D. Van Nostrand Company, Inc., 1958.

Veltman, M., *Facts and Mysteries in Elementary Particle Physics*, World Scientific, New Jersey, 2003.

Venable, W.M., *The Interpretation of Spectra*, Reinhold Publishing Corporation, New York, 1948.

Volk, G., "Toroids, Vortices, Knots, Topology and Quanta," Proceedings of of the 18th Annual Conference the NPA, 6-9 at the University Maryland College Park, MD, Vol. 8, July 2011.

Vrooman, J.R., *Rene Descartes - A Biography*, G.P. Putnam's Sons, New York, 1970.

W

Wagner, O.E., "Structure in the Vacuum," *Frontier Perspectives*, Vol. 10, No. 2, Fall 2001.

Walker, F.L., "The Fluid Space Vortex: Universal Prime Mover," *Physics Essays*, Vol. 15, No. 2, 2002.

Walker, M.S., *Quantum Fuzz – The Strange True Makeup of Everything Around Us*, Prometheus Books, New York, 2017.

Wallace, D.F., *Everything and More - A Compact History of ∞,* W.W. Norton & Company, New York, London, 2003.

Watson, J.D., *The Double Helix*, W.W. Norton & Company, New York, 1980.

Weber, C.S., "VRML Gallery of Fullerenes," Fullerene Library, JSV1.08, 1999.

Weinberg, S., *Dreams of a Final Theory*, Pantheon Books, New York, 1992.

Weir, S.T., Mitchell, A.C. and Nellis, W.J., "Metallization of Fluid Molecular Hydrogen," *Physics Review Letters 76,* 1860, 1996.

Westfall, R., *The Life of Isaac Newton*, Cambridge University Press, New York, NY, 1994.

Wheeler, J.A., *Geons, Black Holes & Quantum Foam – A Life in Physics,* W.W. Norton & Company, Inc., New York, 1998.

Wheeler, J.A., *Geometrodynamics, Topics of Modern Physics, Vol.1*, Academic Press Inc., New York, NY, 1962.

White, H.E., *Introduction to Atomic Spectra*, McGraw-Hill Book Company, Inc., New York, 1934.

Whitney, S.K., "9 Editor's Essays," *Galilean Electrodynamics*, Vol. 16, Special Issue 3, Winter 2005.

Whitney, S.K., *Algebraic Chemistry – Applications and Origins,* Nova Science Publishers, Inc., New York, 2013.

Wiener, N., *Cybernetics or Control and Communication in the Animal and the Machine*, 2nd edition, The MIT Press and John Wiley & Sons, Inc., New York, 1961.

Wigner, E. and Huntington, H.B., "On the Possibility of a Metallic Modification of Hydrogen, *Journal of Chemical Physics* **3** (12): 764, 1935

Wilczek, F., *Longing for the Harmonies – Themes and Variations from Modern Physics*, W.W. Norton & Company, New York, 1987.

Wilczek, F., *The Lightness of Being – Mass, Ether, and the Unification of Forces*, Basic Books, New York, 2008.

Wilczek, F., *A Beautiful Question – Finding Nature's Deep Design*, Penguin Press, New York, 2015.

Williamson, J.G, and van der Mark, M.B., "Is the electron a Photon with Toroidal Topology?," Annales de la Fondation Louis de Broglie, Volume 22, No. 2, 133, (1997).

Witten, E., "Perturbative Gauge Theory as a String Theory in Twistor Space," (2004) (http://arxiv.org/abs/hep-th/0312171)" *Commun Math. Phys.* 252: 189-258.

Woit, P., *Not Even Wrong: The Failure of String Theory and the Search for Unity in Physical Law*, Basic Books, 2006.

Wolff, M., "Origin of the Mysterious Instantaneous Transmission of Events in Science," *The Cosmic Light*, Vol. 4, No. 2, Spring 2002.

Wolfram, S., *A New Kind of Science*, Wolfram Media, LLC, Champaign, IL, 2002.

Wong, H.-S. P., *Carbon Nanotubes and Graphene Device,* Cambridge University Press, Cambridge, UK, 2011.

Z

Zeman, R.K. et al., *Helical/Spiral CT - A Practical Approach*, McGraw-Hill, Inc., New York, 1995.

Zombeck, M.V., *Handbook of Space Astronomy and Astrophysics*, Cambridge University Press, Cambridge, UK, 1990.

Zwikker, C., The Advanced Geometry of Plane Curves and Their Applications, Dover Publications, Inc., New York, 1994.

AUTHOR'S PUBLICATIONS RELATED TO THIS BOOK

1. BOOKS

1. Ginzburg, V.B., *Spiral Grain of the Universe - In Search of the Archimedes File*, University Editions, Inc., Huntington, WV, 1996.

2. Ginzburg, V.B., *Unified Spiral Field and Matter - A Story of a Great Discovery*, Helicola Press, Pittsburgh, PA, 1999.

3. Ginzburg, V.B., *The Unified Spiral Nature of the Quantum & Relativistic Universe*, First edition, Helicola Press, Pittsburgh, PA, 2002.

4. Ginzburg, V.B., *The Unification of Strong, Gravitational & Electric Forces*, Helicola Press, Pittsburgh, PA, 2003.

5. Ginzburg, V.B., *Prime Elements of Ordinary Matter, Dark Matter & Dark Energy*, Helicola Press, Pittsburgh, PA, 2006.

6. Ginzburg, V.B., *Prime Elements of Ordinary Matter, Dark Matter & Dark Energy – Beyond Standard Model & String Theory*, Second revised edition, Universal Publishers, Boca Raton, Florida, 2007.

7. Ginzburg, V.B., *The Spacetime Origin of the Universe*, First edition, Helicola Press, Pittsburgh, PA, 2013.

8. Ginzburg, V.B., *The Spacetime Origin of the Universe with Visible Dark Matter & Energy*, Third edition, Helicola Press, Pittsburgh, PA, 2016.

2. PAPERS

9. Ginzburg, V.B., "Toroidal Spiral Field Theory," *Speculations in Science and Technology* 19 (3) (1996), 165-173.
10. Ginzburg, V.B., "Structure of Atoms and Fields," *Speculations in Science and Technology* 20 (1), (1997), 51-64.
11. Ginzburg, V.B., "Double Helical and Double Toroidal Spiral Fields," *Speculations in Science and Technology*, 21 (2) (1998), 79-89.
12. Ginzburg, V.B., "Nuclear Implosion," *Journal of New Energy*, Vol. 3, No. 4, 1999.
13. Ginzburg, V.B., "Dynamic Aether," *Journal of New Energy*, Vol. 6, No. 1, 2001.
14. Ginzburg, V.B., "Electric Nature of Strong Interactions," *Journal of New Energy*, Vol. 7, No. 1, 2003.
15. Ginzburg, V.B., "Unified Spiral Field Theory – A Quiet Revolution in Physics," *VIA-Vision in Action*, Vol. 2, No. 1 & 2, 2004.
16. Ginzburg, V.B., "The Relativistic Torus and Helix as the Prime Elements of Nature," *Proceedings of the Natural Philosophy Alliance*, Vol. 1, No. 1, Spring 2004.
17. Ginzburg, V.B., "The Unification of Forces," *Proceedings of the Natural Philosophy Alliance*, (Paper presented at the 14th Annual Conference of the NPA, the University of Connecticut at Storrs, Connecticut, 21-25 May 2007), 4.
18. Ginzburg, V.B., "The Origin of the Universe, Part 1: Toryces," *Proceedings of the Natural Philosophy Alliance*, (Paper presented at the 17th Annual Conference of the NPA, the California State University, Long Beach, California, 23-26 June 2010), 7.
19. Ginzburg, V.B., "Basic Concept of 3-Dimensional Spiral String Theory (3D-SST)," (Paper presented at the 18th Annual Conference of the NPA, the University of Maryland, College Park, Maryland, 6-9 July 2011), 8.
20. Ginzburg, V.B., "A Novel Method of Modeling of Fundamental Properties of Materials," Contributed papers from *MS&T15 Materials Science & Technology*, Greater Columbus Convention Center, Columbus, Ohio, USA, October 4-8, 2015.
21. Ginzburg, V.B., "A Novel Method of Modeling of Fundamental Properties of Materials," (Paper presented at *MS&T17 Materials Science & Technology*, Lawrence L. Convention Center, Pittsburgh, Pennsylvania, USA, October 8-12, 2017).

Notes